Teaching and Learning of Knot Theory
in School Mathematics

Akio Kawauchi • Tomoko Yanagimoto
Editors

Teaching and Learning of Knot Theory in School Mathematics

 Springer

OMUP

Editors
Akio Kawauchi
Professor
Osaka City University
3-3-138 Sumiyoshi, Sugimoto
Osaka City, Osaka 558-8585
Japan

Tomoko Yanagimoto
Professor
Osaka Kyoiku University
4-698-1 Asahigaoka
Kashiwara, Osaka 582-8582
Japan

ISBN 978-4-431-54137-0   ISBN 978-4-431-54138-7 (eBook)
DOI 10.1007/978-4-431-54138-7
Springer Tokyo Heidelberg New York Dordrecht London

Library of Congress Control Number: 2012942161

© 2012 by OMUP (Osaka Municipal Universities Press)
1-1, Gakuen-Cho, Naka-Ku, Sakai City, Osaka 599-8531 Japan
TEL: +81-72-251-6533 FAX: +81-72-254-9539

Springer is part of Springer Science+Business Media (www.springer.com)

Foreword

Each level of mathematical knowledge has to be understood as an elaboration of knowledge at a lower level and at the same time as a steppingstone to knowledge at a higher level. This genetic principle applies to both the historical development of mathematics as a scientific discipline and to the learning of individuals. Over centuries the elementary parts of arithmetic, algebra, geometry and analysis have served as the iron basis of mathematical curricula. The enormous progress of the mathematical sciences in the past decades has shown the need to broaden the elementary layer in favor of new topics. Stochastics is the most prominent example. Its elementary parts are nowadays taught already at the primary level. However, even the classical parts of elementary mathematics, in particular geometry, cannot remain untouched. In the past elementary geometry was restricted to Euclidean geometry and to the analytical treatment of lines and planes. Here a broadening of the scope towards "curved objects" is badly needed.

The book is the result of a joint venture of Professor Akio Kawauchi, Osaka City University, well-known for his research in knot theory and the Osaka study group of mathematics education founded by Professor Hirokazu Okamori and now chaired by his successor Professor Tomoko Yanagimoto, Osaka Kyoiku University.

The seven chapters address the teaching and learning of knot theory from several sides. The reader will find an admirably clear and concise introduction into the elementary parts of knot theory, a survey of curricular developments in Japan, and in particular a series of teaching experiments at all levels which not only demonstrate the creativity and the professional expertise of the members of the study group, but also give a lively impression of students' learning processes. In addition the reports show that elementary knot theory is not just a preparation for advanced knot theory but also an excellent means to develop spatial thinking.

The book can be highly recommended for several reasons:
First of all, and that is the main intention of the book, it serves as a comprehensive text for teaching and learning knot theory. Moreover it provides a model for a cooperation of mathematicians and mathematics educators *based on substantial mathematics*. And

finally it is a thorough introduction into the Japanese art of lesson studies – *again in the context of substantial mathematics*.

E. Ch. Wittmann

Erich Ch. Wittmann
Technical University of Dortmund
Faculty of Mathematics
Germany

Preface

It was in July, 2004 that a project to study 'Teaching Knot Theory in Mathematics Education' was formed, which was made possible only when Akio Kawauchi of Osaka City University and Hirokazu Okamori, a professor emeritus of Osaka Kyoiku University met together. Okamori is a researcher of mathematics education. Kawauchi is an expert of knot theory and then he was the leader of "Constitution of wide-angle mathematical basis focused on knots," one of the 21st Century Center of Excellence Programs in Japan from April 2003 to March 2008.

In recent years, knot theory has been making a rapid progress and contributed to elucidating problems of various fields as a science. Knot theory experts are hoping to get across its fun and importance widely to pupils and students in their early years. Also, our cherished hope is that by so doing, we will pick out students with high levels of interest and learning ability in knot theory and bring them up to be excellent experts on it. Experts of mathematics education have a desire to keep developing teaching contents from much broader perspectives beyond the bounds of traditional school mathematics in order to cultivate children's understanding of mathematics.

In pursuit of the realization of the hopes from these two positions, the research project with a team work of researchers of mathematics and mathematics education in university and elementary, junior and senior high school teachers of mathematics was formed by Tomoko Yanagimoto and started discussions of introducing knot theory to educations. All the members began to tackle such work by learning the basics of knot theory. At the same time, we delved into its educational meanings from the viewpoint of mathematics education. Elementary, junior and senior high school teachers held their respective meeting at the school levels and developed teaching materials along with the researchers of mathematics education. Also, based on this, all the members had meetings once a month, where they discussed from their own standpoints under mutual respect as equals and confirmed the systematization of the contents that can be used as materials. We proceeded to discuss the exploration into teaching contents based on the following points.
- Making teaching knot theory something meaningful for pupils rather than something that just propagates knowledge.
- Grasping the real situations of their understanding in their respective years to this

end.

· Developing effective teaching tools as needed.

The report of our research results has already been compiled in Japanese into three issues, attracting the attention of mathematicians, researchers of mathematics education, teacher of mathematics in Japan. In 2006, our research results on knot theory with senior high school students among the SSH (Super Science High School) program, which was one of our experimental practices, won the highest award in the four "first prizes" of the sections of mathematics, physics, chemistry and biology in Japan. The ex post facto evaluation as a part of the 21st Century Center of Excellence Programs of 2008 was of the highest. As one of the reasons for that evaluation, it was pointed out that we had been steadily educating elementary, junior and senior high school students on knot theory.   Further, researchers of representing mathematics educations not only in Japan but also in foreign countries showed interests in our research results. In November 2004, Erich Ch. Wittmann, professor of Dortmund University in Germany, came to visit Osaka and inspected our experimental classwork on mathematics educations of knot theory conducted at Tennoji Junior High School and Tennoji Elementary School attached to Osaka Kyoiku University and showed a deep interest in it. Many times we received encouragements and letters of the research interchanges from Heinrich Besuden, professor of Oldenburg University in Germany and a former chairman of German Mathematics Education Society. Also, whenever we made presentations of our research on teaching knot theory at the conferences of mathematics education in Japan and abroad, including ICME(International Conference on Mathematic Education), lots of interests were directed to our teaching contents, children's reactions, the process developing teaching materials, and the way of proceeding a research project. Thus, this time, we have come to decide to publish our work as a book in English so that we can introduce our research results and its process to many people, including the researchers of mathematics education in the world, knot theory researchers, school teachers of mathematics who have shown interests in our activity.

This book contains an introduction of knot theory, up-to-date topics on it, the educational meanings of knot theory from the perspective of the evolution of mathematics education, the methodology for making knot theory into a curriculum and actual examples of researches and practices, conducted at elementary, junior and senior high schools and a university general education. All the researches and practices have actually been conducted. This book introduces our guidance of knot theory that takes various forms of not only classwork but also experimental teachings with several students, club activity guidance in concerted efforts between senior high schools and universities, and others.

This work was supported in part by the Priority Research of Osaka City University "Mathematics of knots and wide-angle evolutions to scientific objects" and the JSPS Grant-in-Aid for Scientific Research (A) No. 21244005, "Studies of Knot Theory".

Early spring of 2011
Akio Kawauchi and Tomoko Yanagimoto

Notes on the school systems in Japan

In this book, education practices in elementary, junior and senior high schools and university are introduced. Here, we provide a brief description of the school system in Japan which is related to the education practices in order to be understood the situation of the practices well.

Japanese school year starts in April. Therefore, children who are 6 years old when the school year started become first graders in elementary school. Children must attend 9 years of compulsory education from age 6 to 15. Elementary school lasts 6 years and lower secondary school which we call "junior high school" lasts 3 years. Children who are 16 years old in April and pass the entrance examination can enter senior high school. It lasts 3 years. In order to enter private school or school attached to University, children need to pass the entrance examination.

1 class (= 1 school hour) means 45 minutes for elementary school, 50 minutes for junior and senior high schools, 90 minutes for university and college. On the other hand, in the case of club activity, 1hour means 60 minutes.

Authors

Toshihiro Homma (Emeritus Professor of Kobe Shinwa Womens University) 7

Ken-ichi Iwase (Tennoji Junior and Senior High School attached to Osaka Kyoiku
    University) 5.1, 6.3

Akeshi Kamae (Former Minami Junior High School, Osaka City) 5.2

Hiroshi Kanaya(Seifu Senior High School).................................... 6.1, 6.2

Ryohei Kaneda (Suminoe Elementary School, Osaka City) 4

Akio Kawauchi (Osaka City University) 1

Madoka Koyama (Kishinosato Elementary School, Osaka City) 4

Munehisa Matsumoto (Senior High School attached to Osaka Gakuin University)
                                                              .... 6.1, 6.2

Moe Miyazaki (Fijishiro Elementary School, Kyoto City) 4

Naoyuki Masuda (Kansai Soka Senior High School) 6.1, 6.2

Hirokazu Okamori (Emeritus Professor of Osaka Kyoiku University)................2

Yuki Seo, (Shibaura Institute of Technology) 6.4

Hiroji Shibamoto (Hamadera Junior High School, Sakai City) 5.1

Bunji Terada (Former Fukuizumi Junior High School, Sakai City) 5.2

Mikiharu Terada (Shitennoji University) 6.1, 6.2

Hideo Tsuchida (Shitennoji High School) 6.5

MasaruYamamoto (Ishikiri Higashi Elementary School, Higashi Osaka City) 4

Tomoko Yanagimoto (Osaka Kyoiku University) 3,4

Contents

1. What Is Knot Theory?   Why Is It In Mathematics?

In this chapter, we briefly explain some elementary foundations of knot theory. In 1.1, we explain about knots, links and spatial graphs together with several scientific examples. In 1.2, we discuss diagrams of knots, links and spatial graphs and equivalences on knots, links and spatial graphs. Basic problems on knot theory are also explained there. In 1.3, a brief history on knot theory is stated. In 1.4, we explain how the first non-trivial knot is confirmed. In 1.5, the linking number useful to confirm a non-trivial link and the linking degree which is the absolute value of the linking number are explained. In particular, we show that the linking degree is defined directly from an unoriented link. In 1.6, some concluding remarks on this chapter are given. In 1.7, some books on knot theory are listed as general references.

1.1 Knots, links, and spatial graphs

A *knot* is a tangled string in Euclidean 3-space R^3 which is usually considered as a closed tangled string in R^3, and a *link* is the union of some mutually disjoint knots (see Fig.1.1). The AYATORI game (= Cat's cradle play) let us know that a given knot can be deformed into various forms and we feel that it is a difficult problem to judge whether any given two knots are actually the same knot or not.

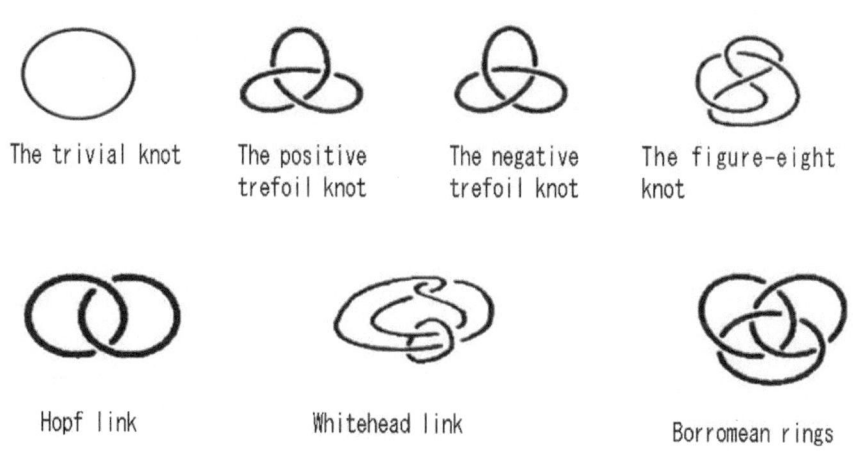

| The trivial knot | The positive trefoil knot | The negative trefoil knot | The figure-eight knot |

| Hopf link | Whitehead link | Borromean rings |

Fig.1.1: Knots and links

A *spatial graph* is the union Γ of some strings with endpoints in R^3 which are mutually disjoint except the endpoints. A *vertex* of a spatial graph Γ is a point in Γ gathering more than 3 strings and an *edge* of Γ is a connected component of the strings obtained from Γ by cutting along all the vertices of Γ. For example, Kinoshita's θ-curve in Fig.1.2 is a spatial graph with two vertices and three edges.

Fig.1.2: Kinoshita's θ-curve

To find out a knot or a link phenomenon in natural science, at first a setting of an object which one can consider as a string becomes important. Here, we show some examples about such objects.

Example 1: It is possible to think a chain (see Fig.1.3) as one string if roughly seeing, but also as a link which are twined round one after another like a string if a little more minutely seeing.

Fig.1.3: A chain

Example 2: A *3-braid* is a knitting of three strings (see, for example, Fig.1.4). Since this knitting pattern is known to be used in a pottery of the JOMON period (an ancient time) in Japan, we see that the people of the JOMON period might understand the fact that the 3-braid is a technology of making a long, strong string like a rope from some short strings like straws. Then, by joining a and a', and b and b' in Fig.1.4, and then by deforming it (without changing near the ends), we can make AWABI MUSUBI (= an abalone knot) of "MIZUHIKI" in Fig.1.5 used for the custom of the present in Japan from ancient times.   Also, although it is a little more difficult, we can also make the same knot by joining b and b', and c and c' in Fig.1.4, and then by deforming it (without changing near the ends). In this way, the knot is also an interested study object for cultural anthropology.

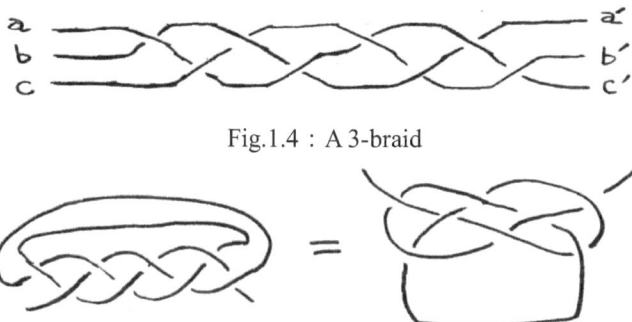

Fig.1.4 : A 3-braid

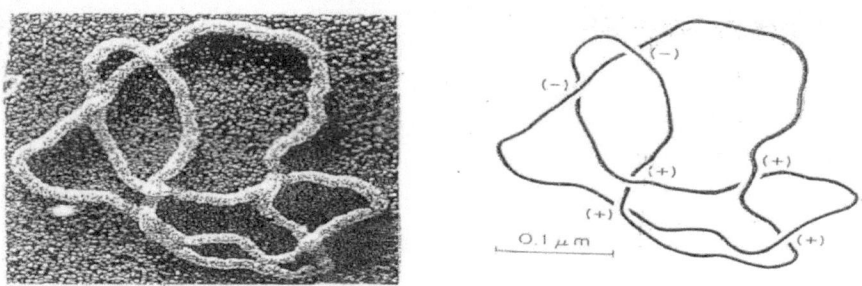

Fig.1.5: AWABI MUSUBI(=an abalone knot) of "MIZUHIKI"

Example 3 : Assume that there are n particles moving without colliding in the plane with a time parameter. Then the track of the motion forms an *n-braid* in the three-dimensional space which is the product space of the plane and the time axis. In the case of 3 particles, we have a 3-braid like the example in Fig.1.4. Every knot and link can cause from an n-braid by taking the closure.

Example 4 : When we consider DNA as one string long rope, there is one which becomes a closed curve called a *DNA knot* (see Fig.1.6).

Fig.1.6: A DNA knot (Acknowledgement :Professor N. R. Cozzarelli)[1]

Example 5 : A *molecular graph* on a molecule in chemistry is a spatial graph whose vertices correspond to the atoms in the molecule and whose edges express the combination data between the atoms by bonds. Topology of molecular graphs has begun to attract attention in researches of the synthetic-chemistry.

---

[1] S. A. Wassermann, J. M. Dungan, N. R. Cozzarelli, Science 229 (1985), 171-174.

4

Example 6 : A protein molecule is considered to be a string consisting of amino acid bases. Some protein molecules such as a prion protein of the Creutzfeldt-Jacob disease and an amyloid β protein of the Alzheimer's disease appear to be more or less related to knot theory.

Example 7 : A large scale structure on the cosmos is recently known to the astrophysicists.

Example 8: The seismometer is a machine which draws the track of an earthquake-motion of an observation point as a spatial curve, called an *earthquake curve* with the time parameter. The analysis of this earthquake curve can be considered as knot theory in the wide sense.

1.2   Diagrams and equivalence on knots, links, and spatial graphs

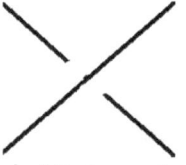

Fig.1.7: A crossing

A knot is in the three-dimensional space and we think that it is made of a very thin string. We present it by a plane curve with only double crossings as they are shown in Fig.1.7, which we call a *knot diagram* or simply a *diagram*. For a link, it is similarly presented and called a *link diagram* or simply a *diagram*. By knots and links, we mean their diagrams unless making confusion. A spatial graph is also presented to the plane with only double points on the edges which we call a *spatial graph diagram* (see Fig.1.3). For two knots, we say that they are the *same knot* or *equivalent knots* if we can deform them into the same shape in the manner of AYATORI game (= Cat's cradle play), i.e., by a finite number of Reidemeister moves I-III in Fig.1.8. For two links, we say similarly that they are the *same link* or *equivalent links* if we can deform them into the same shape in the manner of AYATORI game, namely by a finite number of Reidemeister moves I-III. For two spatial graphs, we say that they are the *same graph* or *equivalent graphs* if we can deform them into the same shape in the manner of AYATORI game or in other word by a finite number of Reidemeister moves I-V in Fig.1.8. A knot is called a *trivial knot* if it is equivalent to a circle in the plane like a rubber band. Also, a link is called a *trivial link* if it is equivalent to a union of separated trivial knots.

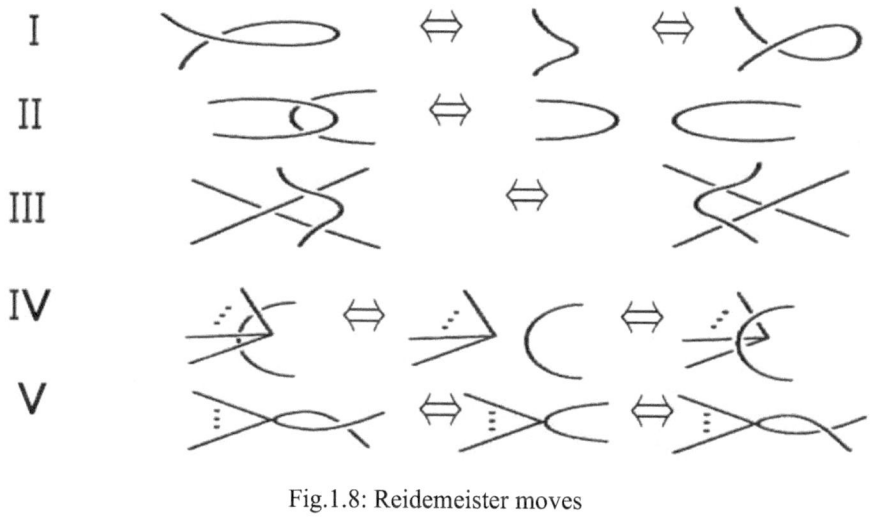

Fig.1.8: Reidemeister moves

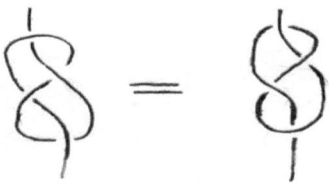

Fig.1.9: The same arc knot

In daily life, we normally think of an arc knot in Fig.1. 9. At that point, we consider that the end points extend virtually long and endless. If we can deform them into the same shape in the manner of AYATORI game without moving the parts that are extended, then they are considered to be the same knot. A link with two end points can be considered similarly to the case of an arc knot (See Fig.1.10). If a link has more than two end points, then such a link no longer has any meaning unless the data extending the end points are definitely given.

Fig.1.10: A link with two end points

The main purpose of knot theory is to solve the following two problems (which are related to each other):

Equivalence Problem : Given two knots (or links, spatial graphs), determine whether or not they are equivalent.

Classification Problem : Enumerate all the knots (or links, spatial graphs) up to equivalences.

To solve these problems, it is very important to develop *topological invariants*, (namely, quantities which are invariant under the Reidemeister moves) of knots, links and spatial graphs.

1.3 A brief history of knot theory

About the beginning of a scientific study of knot theory, an opinion is convincing which is associated with the atomic theory of vortex atoms in ether around the end of the nineteenth century, though it is can be traced back to a note by J. B. Listing, a disciple of C. F. Gauss in 1849. In this note, it has been that the mirror image of the figure-eight knot is the same knot (Fig.1.11). Up until the 1930s, important researches were made by K. Reidemeister and H. Seifert in Germany and J. W. Alexander in U.S.A., etc. From 1940 until the 1970's, one may say that basic mathematical theory on knot theory was established with R. H. Fox in U.S.A. as a center. In Japan, from around 1960, H. Terasaka, T. Homma, S. Kinoshita (later, moving to U.S.A.) and K. Murasugi (later, moving to Canada), F. Hosokawa, etc. have begun to make contributions to knot theory. From around 1980, knot theory came to attention not only in almost all areas of mathematics, but also in the fields of science that will be cutting-edge researches, such as gene synthesis, quantum statistical mechanics, soft matter physics, biochemistry, polymer network, applied chemistry···. The international conference "Knot theory and related topics" received the world's first was held at Osaka as a satellite conference of ICM Kyoto in 1990, from whose proceedings "Knots 90" (A. Kawauchi, e.d., Walter de Gruyter, 1992) one may feel a fever of an expansion of knot theory.

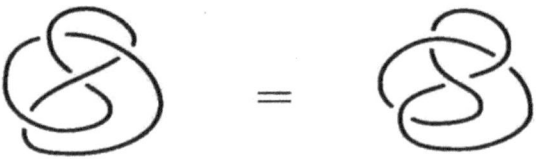

Fig.1.11: Equivalence of the figure-eight knot and the mirror image

1.4   The first non-trivial knots

Of knots and related concerns that are normally used casually in everyday life, let me say here the simplest mathematical proof that there is a non-trivial knot. (This proof uses an argument of the tri-colorability of a knot which is well-known to the experts of knot theory as the representation theory of the knot group into the dihedral group of order 6.) Given a knot diagram, we color all the edges connecting the crossings by using three colors (e.g., red, blue and yellow) by imposing in the vicinity of every crossing the condition that we color the upper arc by the same color and color the lower two edges by the same color or different colors so that one color or three colors are used in the vicinity of every crossing. By this method, we can always color every knot diagram by one color. For some knot diagrams, we can also color them by tri colors (see Fig.1.12). Such a knot is called a *tri-colorable knot.*

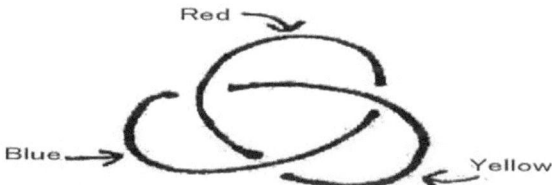

Fig.1.12: The trefoil knot is tri-colorable

Theorem :   Every tri-colorable knot is a non-trivial knot.

The reason why this is true is because we can easily check that any knot diagram D' transformed from a tri-colorable knot diagram D by Reidemeister moves I, II, III (see Fig.1.8) is also tri-colorable. We recommend to confirm this fact for various knots.

1.5   Understanding the linking number

The linking number of a link of two oriented knot components is the most fundamental topological invariant in knot theory. However, because it takes a value in the integers, this invariant cannot de defined without the notion of a negative integer. In this section, we first introduce the linking number. Next, we introduce the linking degree, the absolute value of the linking number, which we can define and compute directly from a link diagram without using the notion of negative integers. Thus, we can suggest a method that we introduce first the linking degree and then bring the linking number forward while we introduce the notion of negative integers.

8

Fig.1.13: A link diagram

1.5.1. Linking number   We shall explain how to define the linking number of an oriented link. We consider a link L of two oriented knot components as in Fig.1.13. Every crossing between the distinct knot components coincides with one of the four crossings shown in Fig.1.14, where the signs of two crossings in the left hand side are defined to be +1 and the signs of two crossings in the right hand side are defined to be -1. Let m be the sum of signs of all the crossings between the distinct knot components. We see that this integer m is always even and does not change under Reidemeister moves I, II, III   (see Fig.1.8), namely m is a topological invariant. Then we define the *linking number* v of the link L to be the integer $m/2$. When we reverse the orientation of one component oL, the sign of every crossing between the distinct components is changed so that the linking number v of L is changed into $-v$. For example, the link diagram in Fig.1.13 has the two (+1)-crossings and the four (-1)-crossings and the linking number v is given by

$$v = (2 - 4) / 2 = -1.$$

$$+1 \qquad\qquad +1 \qquad\qquad\qquad -1 \qquad\qquad -1$$

Fig.1.14: The sign of a crossing

1.5.2. Linking degree   We consider a diagram of an unoriented link L with knot components J and K as in Fig.1.15. To distinguish between J and K, we denote K by a bold line. We attach an orientation to one of the components J and K, say J as shown in Fig.1.16. Every crossing between the oriented component J and the unoriented component K coincides with one of the two crossings in Fig.1.17. We construct a *meridian loop*, an oriented loop surrounding K one time in every crossing between J and K as in Fig.1.18. For example, we obtain the left-sided diagram in Fig.1.19 from

Fig.1.16, from which we obtain the right-sided diagram in Fig.1.19 by sliding these meridian loops along K.

Fig.1.15: An unoriented diagram

Fig.1.16: A link diagram with the component J oriented

Fig.1.17: The situations of a crossing

We denote by (P, Q) a pair of the subsets of loops with the same orientations in the set of meridian loops around K obtained from the oriented knot J. Let p and q be the numbers of the elements in P and Q, respectively, where we take $p \leqq q$. Then let $n = q - p$, and the *linking degree* d of a link L consisting of the unoriented knot components J and K is defined by

$$d = n/2 = (q-p)/2.$$

For example, the linking degree d of the link L in Fig.1.15 is computed from Fig.1.19 to be

$$d = (4-2)/2 = 1.$$

If we take the reversed orientation on J, then the orientations of all the meridian loops are reversed and the non-negative difference n between the numbers of the meridian loops with the same orientations is unchanged (though the roles of p and q are interchanged), so that the linking degree d is independent of a choice of any orientation on J.

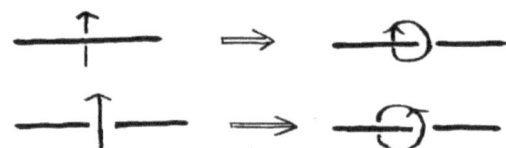

Fig.1.18: Constructing a meridian loop of K

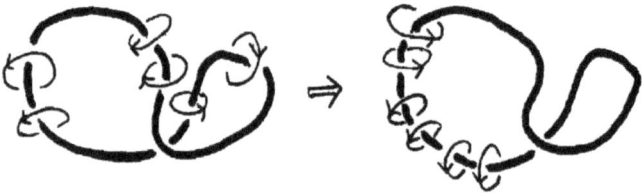

Fig.1.19

We show that the same number n is obtained as the non-negative difference of the numbers of the meridian loops with the same orientations even if we consider the meridian loops around J obtained from the component K oriented in any direction (instead of J). In fact, if we consider J and K as oriented knots, then we see that the orientation of the meridian loop is locally determined as it is shown in Fig.1.20, so that a pair of the numbers of the meridian loops around J with the same orientations obtained from the oriented knot K coincides with the pair (up to ordering) of the numbers of the meridian loops around K with the same orientations obtained from the oriented knot J. This implies that the linking degree d of a link L is a non-negative rational number with denominator 2 which is independent of choices of the components and the orientations of J, K.

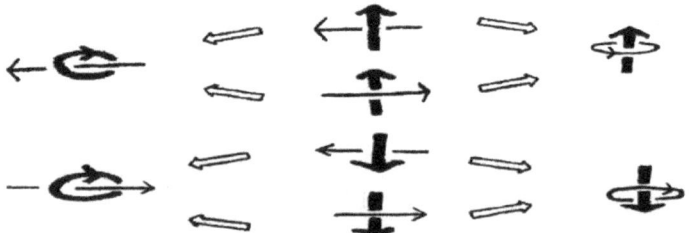

Fig.1.20: The orientation of a meridian loop is locally determined

We show the following (1) and (2).

(1) The linking degree d is unchanged under Reidemeister moves I, II, III and hence it is a topological invariant.

(2) The linking degree d takes a value in the natural numbers or zero. When we orient a component J, the linking degree d is computed to be the difference of the numbers of the meridian loops with the same orientations around the other component K on the crossings of J which are upper than K, or the difference of the numbers of the meridian loops around K with the same orientations on the crossings of J which are lower than K.

First, we show (1). The proof is made by considering a pair of the numbers of the meridian loops with the same orientations around K obtained from an oriented J. The numbers n and d are unchanged under Reidemeister moves except the moves relating to both J and K. In particular, they are unchanged by the Reidemeister move I. For the Reidemeister move II relating both J and K, we can see from Fig.1.21 that the numbers n, d are unchanged under the Reidemeister move II.

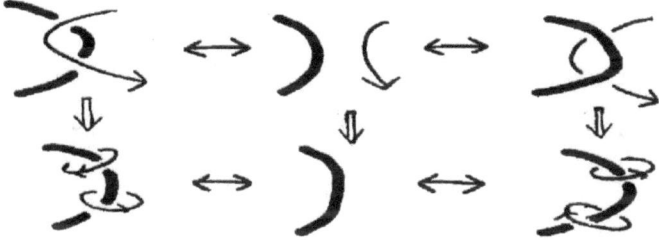

Fig.1.21: Reidemeister move II relating to J and K

For the Reidemeister move III relating both J and K, it is sufficient to consider the cases in the left-hand side of Fig.1.22, if necessary, by changing the roles of J and K. By examining the pictures in the right-hand side of Fig.1.22, we see also that the numbers n, d are unchanged under the Reidemeister move III, showing (1).

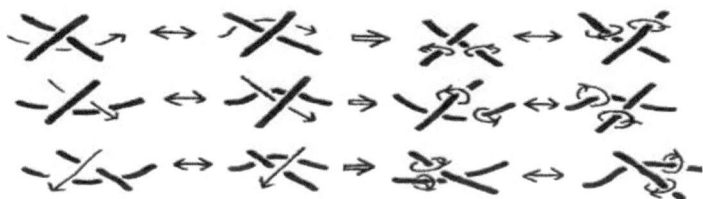

Fig.1.22: Reidemeister move III relating to J and K

Next, we show (2). We can see from Fig.1.20 and the definitions of the linking number and the linking degree that the linking degree d is the absolute value of the linking number v. Therefore, if we assume that the linking number v takes a value in the integers, we can see that the linking degree d takes a value in the natural numbers or zero. Here, we give a direct proof about it. Let (P, Q) be a pair of the subsets of loops with the same orientations in the set of meridian loops around K obtained from an oriented knot J. Let E and F be the subsets of P obtained from the crossings of J upper and lower than K, respectively. Let e and f be the numbers of the elements of E and F, respectively. Also, let G and H be the subsets of Q obtained from the crossings of J upper and lower than K, respectively, and g and f the numbers of the elements of G and H, respectively. Then for the numbers p, q of the elements of P and Q, where we assume $p \leq q$, we have

$$p=e+f, \qquad q=g+h.$$

We change all the lower crossings of J into the upper crossings of J, so that all the crossings of J are upper than K by a crossing change operation in Fig.1.23. By this change, the numbers $p=e+f$ and $q=g+h$ of the meridian loops with the same orientations are changed into $e+h$ and $g+f$, respectively (cf. Fig.1.18). Since now the knot diagram of J is upper than the diagram of K, we can move the knot diagrams of J and K by Reidemeister moves II, III so that they do not meet. By the topological invariance of the linking degree, we have

$$(e+h) - (g+f) = 0, \text{ namely } g-e = h-f.$$

Hence, we have

$$d = n/2 = (q-p)/2 = g-e = h-f$$

and (2) is proved.

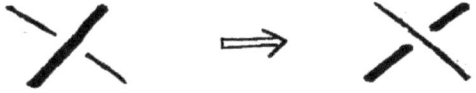

Fig.1.23: A crossing-change operation

1.5.3. Computing examples of the linking degree using the meridian loops on the upper crossings

Here are some examples of computations on the linking degree.

Example 0. We introduce an orientation on J as in Fig.1.16 to compute the linking degree of a link in Fig.1.15. Then the meridian loops on the upper crossings belonging to J are the loops in Fig.1.24. Thus, we have $d = 2\text{-}1 = 1$.

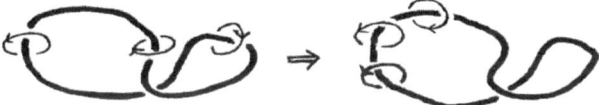

Fig.1.24: The meridian loops on the upper crossings of the diagram of Fig.1.16

Example 1 .   The linking degree d of the Hopf link in Fig.1.25 is computed to be $d = 1$.

Fig.1.25: A computing process for the Hopf link

Example 2.   The linking degree d of the Whitehead link in Fig.1.26 is computed to be $d = 1\text{-}1 = 0$.

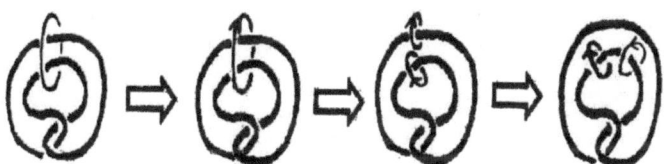

Fig.1.26: A computing process for the Whitehead link

Example 3.   The linking degree d of a 2-braid link in Fig.1.27 is computed to be $d = 2$.

Fig.1.27: A computing process for a 2-braid link

Example 4.   The linking degree d of the parallel link of a trefoil knot in Fig.1.28 is computed to be $d = 3$.

Fig.1.28: A computing process for the parallel link of a trefoil knot

Remark. The linking degree d of a twisted parallel link of a trefoil knot in Fig.1.29 is computed to be $d = 3\text{-}3 = 0$.

Fig.1.29:   A twisted parallel link of a trefoil knot

Example 5. The linking degree d of a parallel link of the figure-eight knot in Fig.1.30 is computed to be $d = 2\text{-}2 = 0$.

Fig.1.30: A computing process for a parallel link of the figure-eight knot

1.6   Conclusion

It is an ultimate purpose of knot theory to clarify a topological difference of knot phenomena in mathematics and in science. In this study, a building power and a computational ability in mathematics are needed in addition to the intuition power having to do with a figure. We can watch a knot with eyes and our ability of space perception will be grown up by playing with it. Knot theory is a subject suitable for understanding nature deeply and desirable for learning in an early age. Concerning the linking number, we suggested here a method that we introduce first the linking degree and then bring the linking number forward while we introduce the notion of negative integers.

1.7 References

There are many books on knot theory some of which are listed here. Among them, the first appearing book is "Knotentheorie" written in German in 1933 by K. Reidemeister.

[1] C. C. Adams, The Knot Book, American Mathematical Society, 2004.

[2] G. Burde, H. Zieschang, Knots, de Gruyter, 1986.

[3] P. R. Cromwell, Knots and Links, Dover Publications, 2008.

[4] R. H. Crowell & R. H. Fox, Introduction to Knot Theory, Springer-Verlag, 1977 (This book was first appeared from Ginn and Co. in 1963).

[5] L. H. Kauffman, Knots and Physics, World Scientific, 1991.

[6] L. H. Kauffman, Formal Knot Theory, Dover Publications, 2006 (This book was first appeared from American Mathematical Society in 1983).

[7] A. Kawauchi, A Survey of Knot Theory, Birkhauser Verlag, 1996 (This book is an expanded version of the book "Knot Theory" published in Japanese from Springer Verlag, Tokyo in 1990).

[8] A. Kawauchi, Lectures on Knot Theory (in Japanese), KYORITSU SHUPPAN CO., LTD, 2007

[9] W. B. R. Lickorish, An Introduction to Knot Theory, Springer Verlag, 1997.

[10] C. Livingston, Knot Theory, Mathematical Association of America Textbooks, 1996.

[11] V. Manturov, Knot Theory, Chapman & Hall/CRC, 2004.

[12] K. Murasugi, Knot Theory and Its Applications, Birkhauser Verlag, 2004 (This book was first appeared as a Japanese book with the same title from Nippon Hyoronsha Co. Ltd. Publishers in 1993).

[13] K. Reidemeister, Knotentheorie, Ergebn. Math. Grnnzgeb. 1, Springer Verlag, 1933.

[14] D. Rolfsen, Knots and Links, American Mathematical Society, 2003 (This book was first appeared from Publish or Perish, Inc. in 1976).

2. The Evolution of Mathematics Education
-forwarding the research and practice of teaching knot theory in mathematics education-

2.1 Overview

Our project team has compiled the report 'Teaching Knot Theory in Mathematics Education' written in Japanese into three issues since 2005. Luckily, the researchers representing mathematics educations in Japan gave us their comments in response to this report. We take them as instructions and encouragements to our project. We introduce some of them here first.

Next, referring to the historical developments of arithmetic and mathematics education in our country, we state what has motivated us to tackle the challenge of 'teaching knot theory in mathematics education. '

2.2 Comments to our second issue

H. Fujita who was the chairman of the Mathematics Education Society

"I am appreciative of your second issue of the report, 'An approach to Teaching Knot Theory –in elementary, junior/senior, high schools' sent to me this time.

How remarkable it is that knot theory, based directly on the perspective and methodology of new mathematics, can possibly be practiced in mathematics education for pupils and students alike. The concept and way of thinking underlying in Knot Theory is all the more important because this is part of the frontiers of mathematics, and the fact that this does not need much preparatory practice also appeals to me.

History shows that there are turns and twists before such developments are led into school curriculums. Differential and integral calculus also needed Perry movement and it took as long as two centuries to have it introduced into the secondary education. I should say the theory of probability took much longer.

Patience may be needed in making knot theory into a curriculum, unlike in introducing the computer well responsive to the trend of the world. However, I find your pioneering mathematics education of full significance. Keep up the good work."

S. Okabe who was the chairman of the Society of Practical Research of Mathematics Education

"I've learned a lot through reading your papers, "An Approach to Teaching Knot Theory -in elementary school." Pretending to be an elementary school pupil, I actually

made a knot with a string and tried to move it.

As for 'spatial visualization,' in particular, as you had written, I really have come to realize that what was left unseen can become visible through changing operational forms, such as moving an object, looking at it from the reverse, seeing it from the back of the paper(a reflected image). I find this an epochal practice indeed."

Y. Yamagishi, Researcher of Mathematics Education.

"I thank you very much for your summary of an elaborate practice and research of 'Teaching Knot Theory in Mathematics Education' in cooperation with elementary, junior/senior high school /university. What a laborious work this is! I could see you all working well on this in high spirits. I know it is really tough to sum up contents of new mathematics education, but I do pray and hope that your research result is in a great success.

About twenty years have already passed since I left from teaching. In the meantime, and even before that, various kinds of sciences, including mathematics, have rapidly made progress, which, needless to say, has brought about more or less changes in teaching contents of various kinds. Keeping in mind what kind of arithmetic and mathematics education to aim at, I had not a few opportunities to give lessons with the contents far from textbooks. Since the pupils and students to teach in my charge were still on their way to learning and development, I strained every nerve for them, because it is basically of utmost importance in education to properly grasp their way of thinking, levels of their perception and their comments.

"The practice in elementary school" in the report gave me some food for thought. I thought of a possibility of letting pupils freely make "a knot" and then make it into a seal as regards "the question of turning a knot over" in the post-research. This could look like '花押', Kaou used by farmer feudal lords, rather than "a seal", because I thought you can assign this process to them as a new challenging situation, where each of them makes his or her own thing, so that they will find something more positive and fulfilling in learning through this process, which leads to their deeper understanding of what they have studied. This is my real hope.

I think of the same as for junior/senior high schools. The contents they're learned should depend on their grades, that is, so to speak, closed ones, but as I mentioned before, I would rather you went ahead with your research centering around your ideas since I've left from teaching.

Still, I highly respect your laborious work of having conducted together with ex post facto researchers even advanced researches, whose necessity should arise and whose rounded off meticulous plan, issues and subjects should be addressed in future regarding the teaching contents are lessons you demonstrated.

I thought you can obtain valuable materials from junior school students' views as well as a summary of each chapter."

We introduced the comments of three teachers as mentioned above, which showed us the bravery and expectation in how we should take up new mathematics in mathematics education. We think they pointed out significances in the approach to knot theory in mathematics education by our research project team.

2.3 Historical developments of the arithmetic and mathematics education in our country

Let me pick up very briefly what happened around us in the earliest stages and the modernization period in the process of our mathematics education here.

2.3.1 The Earliest Stages Since 1873 and 1874, there had been government Osaka Teacher's College (later Osaka Prefectural Teacher's College) in Osaka City. Our mathematics education and textbooks at that time were based on translations. K. Ogura said that D. Murray, mathematics professor of Rutgers University in U.S., was invited in 1873 as a school superintendent to the Education Ministry, staying in Japan until the end of 1878. Several times since 1874, he had met with N. Okamoto, mathematics teacher at National Osaka Teacher's College. Murray reported to the Education Ministry in the second annual report of the Education Ministry that Okamoto had a genius for mathematics and that Murray found him to be a hardworking and intelligent mathematics teacher with detailed knowledge of mathematics. Okamoto, mathematician of Japan and learned Western mathematics, was probably highly evaluated his highly sense of western mathematics.

2.3.2 Modernization period   At the end of the 19th century, the Industrial Revolution got doing, and science and technology made rapid advances. Thus, the remodeling movement of mathematics education was urged to get started in our country as well as in Europe and America. The year of 1918 saw 'the conference of teachers of mathematics at teacher's colleges, middle schools, higher girls' schools across the country' (later, morphing into the current Japan Society of Mathematics Education) at Tokyo High Teacher's College.   Let me introduce K. Kitagawa, active in the movement then. He was a teacher in charge of mathematics at the middle school attached to the Tokyo Higher Teacher's College and energetically proposing the introduction of 'solid geometry' into the middle school mathematics education in Japan

20

at the above mentioned conference. At that point of time, he translated P. Treutlein's book 'An Intuitive Space Lesson' into a Japanese book with the title 'An Intuitive Space Lesson for primary school pupils and middle school first-year students'. In the preface of his translation, Kitagawa said emphatically "Sound progress in physics and chemistry must be inevitably followed by improved mathematics education. Physical and chemical science without mathematics cannot possibly move from the qualitative handling to the quantitative theory, weakening its existence as a science. Even if it enjoys the height of flourishing temporarily, it is only a passing phenomenon. From this viewpoint, what has to be done in mathematics education is to make better mathematics education." In P. Treutlein's book, F. Klein having published Erlangen Catalog, epochal suggestions regarding mathematics education had written the preface "..... the chemical bond between intuition and theory.....". This book contains "Eye measurement / observation and imitation of magnitude", "Language practice", "Hard practice", "Considerations of real life" which is to deserve attention now, much more hereafter. In fact, I should also mention here that British traditional geometry education had gotten adopted into our geometry education.

'Modern New Geometry (basics)' by Y. Abe (Tokyo Kaiseikan publication, 1935), which I used in my first, second, third years of middle school, apparently took in what above-mentioned Kitagawa had emphasized. Although the text made little impression on me in those days, its contents must have been quite up-to-date. I still remember now clearly that I made strenuous efforts to make a model of a regular dodecahedron as a home assignment of my first-year geometry class.

Incidentally, then, Kitagawa majored in geophysics at Kyoto University, studied in Europe and America and served as the first president of Osaka University of Arts and Science.

Now, in an age of internationalization or globalization, R. B. Davis, professor of Rutgers University in U.S., W.L.Fischer, professor of Erlangen-Nürnberg University in Germany where F. Klein used to be and our research group have had comings or goings through research interactions over mathematics education for the 21th century with computer education and others centering around our topics since 1980, which I am going to talk about later.

2.3.3 The period of the "modernization of mathematics movement"　　In 1958, when modernization of mathematics education got discussed in Japan, Mathematics Education Society was established, where the following fields were, then, assumed by S. Iyanaga, Y. Akizuki, W. Shibagaki, I. Tajima, K. Kato, K. Yokochi to promote researches of mathematics education as 'study approaches'.

- Psychological approach to mathematics education
- Social scientific approach
- Historical approach to mathematics education
- Educational study of the history of mathematics
- An approach from comparative pedagogy to mathematics education
- Theoretical study of teaching contents
- Theoretical study of teaching curricula
- Mathematical approaches – considering the developments of teaching contents from the perspective of new mathematics

Also, they came up with the following fields as research contents.

- Algebraic field
- Geometrical field
- Analytical field
- Fields of probability and statistics

Furthermore, I stated to take up the following fields from around 1968.

- Education Administrative study of mathematics education
- Brain physiological study – this field is now receiving a lot of attention as a research of brain science, in connection with 'functions of the right hemisphere and the left hemisphere of the brain' and 'lesson involving teaching tools and operations' and 'admonitory lectures'

With the development of the computer, new mathematical fields also have recently been making great strides. Thus, in mathematics education, relations with ICT (Information Communication Technology) teaching, fractal, chaos, fuzzy, and discrete mathematics as post-Bourbaki mathematics have been investigated. Knot theory can be said to be one of the mathematical fields to be discussed in new mathematics. Besides these, theories of lessons, teaching tools and operations, clinical methods are noticed.

I do think that mathematics education is to "evolve" as learning or as a growing organism with the above-mentioned various contents and methods more organically bound together. The state of things, however, is far off.

2.4 International research interactions of mathematics education

While in Japan, mathematics education of various countries came to be known, first, by way of literature research through books, our research group has had occasions for research interchanges through human comings and goings and promoted international research exchange by mutually presenting research results.

In 1980, K. Yokochi, S. Machida and I attended the 4th ICME held at Berkeley and had a presentation on geometry education. Then, H. Besuden, professor of Oldenburg University, sitting in conference, vice chairman of West Germany's Mathematics Education Society, showed his keen interest in our presentation. In November 1981, the following year, Besuden came to Osaka for a research interchange with us.

He talked mainly about these matters.

- · The school system of Germany
- · Teaching curricula of geometry
- · Methods of geometry education

Furthermore, citing the education of isometric transformation (parallel movement, symmetrical movement, rotary movement) by spiral method, Besuden introduced line symmetry and rotation symmetry with square boards of his own devising for the lower grades.

While Besuden made the pupils aware of sight angle, parallel, congruence using already got-up square boards, we led them to mathematically recognize things around them, which could be the difference between us.

On the same day, we talked, in the main, about the following matters.

A. Pupils' spatial visualization and geometry education

- · Geometry education at elementary school through operations
- · Meanings of teaching tools
- · Significance of the use of the computer

B. An experiment in geometry education

- · 'Direction and movement' education for lower grades in elementary school

What's more, the lesson, "Making a stamp -mirror image-" conducted by T. Ishimi (Yanagimoto) for record graders at Tennoji elementary school attached to Osaka Kyoiku University, was later introduced by Besuden in the West German journal, 'Science Mathematics Primary Graders' (1983).

Then, Besuden made his comment on own presentation like this "It surprised me a lot that you demonstrated here what I had never even thought of before, and that Ms. Ishimi's lesson perfectly realized what I've been thinking of so far.

What Okamori and other teachers have done here exactly fits my thoughts, though, of course, those who practice mathematics education all may not think in the same way as Okamori, Yokochi and we do.

Several years ago, I had an opportunity to visit a professor living in the state of Arizona, the U.S. He had felt difficulties with the educational system just as we had been thinking, because, in fact, children never used either teaching tools or their hands. Now then, Arizona University with one organization of mathematics education is thinking of a way of teaching mathematics: Pupils actually complete things using something. I am so pleased to find different people on the other side of the earth thinking of the same as we do and putting that into practice. Since I have no idea what's going on in China and other countries, I hope to see their situations once indeed.

Needless to say, mathematics education, even with its objective closed, ultimately reflects children's thinking of the abstract since they think of more abstract things step by step.

There was a somewhat wrong way of thinking when it comes to 'abstraction'. As a matter of fact, we have been instructing more abstract and fast growing pupils to think. There was a tendency to abstraction or to think of abstraction to be developed at a very early stage in life. Just as J. Fujii said, this is due to a part of the failure of the modernization of mathematics education, I think. Luckily, however, we were led to notice such a mistake.   We have fully noticed the 'abstraction' as a part of the process to find out the abstract ultimately through a part of the process without teaching it from the beginning.

Last year, when we met together on the occasion of the 4th ICME in Berkeley, we did not know each other at all, but here now, I am so pleased to find people who think in the same way as we do and are advancing in the very same direction as ours. I hope with many thanks that this acquaintance with ours may develop into deepening our thoughts and making further investigation together from now on. Similar types of research interchanges were held at Yamanashi University and Saitama University. In May, 1982, the following year, we (Yokochi, Machida and Okamori) were invited by Germany Society of Mathematics Education with the introduction from Besuden to conduct our research interchanges at the 5 universities: Oldenburg University, Münster University, Köln University, Erlangen University, Bremen University

I made a lecture on the title of 'Teaching Materials (Including Computer) and Valuable works in Geometry Education', and during the research interchange, the 'solid dihedral angle cutting machines' constructed at my laboratory for the first time in the world were dedicated to Oldenburg University and Erlangen University.   After this interchange, I got a journal "International Colloquium on Geometry Teaching" sent kindly to me by G. Becker, professor of Bremen University, which wrote about geometry teaching presented and discussed during the research meeting of mathematics education at universities in Europe.

The paper, "First steps in proving by grade pupils" by G. Becker on the chapter of "Objectives in geometry teaching" and the paper, "Geometrie et dependance" by G. Papy on the chapter of "Recent problems and in new trends in geometry trends teaching" and the paper, "Toward relevance geometry" by P. A. Bishop were printed in the journal. I was struck with a part in the Bishop's paper, "break from the Piaget-myth", that is, a bold suggestion.

On this occasion, we founded a 5-nation international conference on mathematics education (China, Germany, U.S., France and Japan) with Yokochi as its chairman and our research interchanges have come to be held every year.

After the ICME7 in July 1992, the 5-nation conference was held at Rutgers University in U.S., where I made a lecture on the title of "Mathematics Teacher Training in Information-Centered Society."

In Rutgers University there was a glorious memory of Davis and we had research interchanges with him since he was a professor of Illinois University days. He came to visit Osaka Kyoiku University, too. When I presented him with "Mathematics Education and Personal Computers" of my own editing at Rutgers University, Davis gave me his book, "Learning Mathematics -The Cognitive Science Approach To Mathematics Education-" as a gift.

Now, Besuden is, even today, tackling mathematics education. He came to know that we have been addressing 'Knot Theory' in mathematics education of late and sent us the following letter in December 2008.

Dear Prof. Okamori,

I was delighted to receive some results of your work on introducing some knot theory in school mathematics in your country. This is just what I also suggest to be done in our schools. I would like to encourage your study group to go on promoting problems in knot theory (based on praxis) to be put in the curriculum. That does help developing spatial abilities of the pupils.

Good luck to your work and to you personally.

Very sincerely

Heinrich Besuden

An approach from comparative pedagogy is among the study approaches to mathematics education. As I mentioned first, it was through literature research based on books that the situation of mathematics education in various foreign countries started to be introduced to Japan, but we have had occasions of research interchanges through human comings and goings and presented our mutual research results and promoted international research interchanges.

We have been positively advancing our study approach in pursuit of the actual conditions of children in school education in Japan. Based on these, we have been conducting research interchanges with researchers of various foreign countries. We have been blessed with on-going suggestions as to knot theory in mathematics education through our research interchanges.

2.5   What knot theory means to mathematics education

S. Iyanaga wrote on 'Mathematics in the 21st century' in the book, "What is to come hereafter to Japan and the world in the 21st century" (edited by the editorial stuff of Iwanami Shoten Co. Ltd. 1983). The finis of the book is read like this, "In the early parts of the 19th and the 20th centuries, K. F. Gauss and D. Hillbert in their younger days and mature age, respectively emerged and led the way to mathematics in the coming centuries. I do hope the same is true of the next century. 'Knot theory in mathematics' is mathematics of post-Bourbaki mathematics and is based on life and culture and is a basis for elucidation and progress of various sciences, such as DNA and others.

Our project team has been conducting demonstrative researches based on educational experiments in order to introduce a field of new mathematics, "Knot theory" into education.   Neither Perry Movement at the time of the modernization of mathematics education (100 years ago) nor Cambridge report at the time of the modern mathematics movement of education (50 years ago) was based on such positive researches.   Thus, I believe that our project team's approach overcomes the dispute at around the time of the modern mathematics movement in Japan over both the research closely connected with N. Bourbaki and J. Piaget and children's understanding-centered research of mathematics education.

3. The Background of Developing Teaching Contents of Knot Theory

In this chapter, we introduce a methodology for making knot theory into a curriculum and actual examples of researches and practices, conducted at elementary, junior and senior high schools and a university general education. In 3.1, we introduce that a recent course of study in Japan has changed a little on the flexibility of constructing a curriculum in each school. In 3.2, we explain a viewpoint to introduce a new mathematics into the education based on considerations of the "modernization of mathematics movement" in 1960-1970's. In 3.3, an approach to create teaching contents of knot theory is introduced based on our project. We show how to organize and proceed the project. This chapter is a detailed version of the paper "The methodology for Creating New Teaching Contents in Mathematics Education" contributing to "Forum on Public Policy", Journal of the Oxford Round Table.

3.1 Recent trend in the course of study and actual situation of schools in Japan

The *course of study in Japan* is discussed together with actual situation of schools especially from the view point of legal binding. The course of study in Japan in recent years has come to be gradually alleviated from legal binding. It is related to changes in the content of education. The course of study in Japan has gained legal binding force since 1958. The following system governs from the creation of the course of study to the use of textbooks. First, the *MEXT (Ministry of Education, Culture, Sports, Science and Technology)* consults the committee of the Minister of Education and Science (previously the Education Process Committee, presently the Central Education Council) about improvements of the criterion of education process before the revision of the course of study. The course of study is prepared based on the report of the committee.

The Elementary and Junior High School Education Bureau and the Physical Education Bureau take charge of and carry out the actual work of the creation of the course of study by organizing the "Course of Study Creation Committee." The finalized documents are publicized in the form of announcement. Several textbook publishers prepare textbooks in accordance with the announcement and apply for the textbook examination by the MEXT. In the textbook examination, contents beyond the scope of the course of study are completely eliminated. Public schools must use textbooks approved by the examination, and virtually most private schools also use the textbooks.

Although the course of study indeed permits each school to arrange appropriate curriculums, they must comply with the laws and the course of study. Therefore, Japanese course of study can be said to have substantial legal binding force.

In its revisions in recent years, however, it has become somewhat more flexible. The trend is related to the following changes in educational contents. According to the educational contents, they have gained flexibility in roughly three measures.

(1) Each school has come to be allowed to arrange its own curriculums in the period for integrated study" a region separate from the conventional courses of study.

(2) Each school has come to be allowed to put advanced contents freely in conventional courses of study.

(3) In order to promote enrichment in the education of technology, science and mathematics, the "*Super Science High School* Project" was established on which the designated schools are encouraged to study the development of curriculums with advanced contents.

Voluntary arrangement of curriculums beyond the limits of the course of study, however, is not so easy for schools who have not been allowed to arrange curriculums on their own for about 50 years. I will show you the current situation especially of mathematics education in elementary and junior high schools for the measures (1) and (2), and in senior high schools for the measure (3).

3.1.1 Curriculum arrangement for "the period of integrated study" Current course of study is the one revised in 2002 aiming at acquiring of basics and developing the zest for living in "the cram-free education".[1],[2] Here the "period of integrated study," namely hours to be spent in learning subjects such as international understanding, information, the environment, welfare, and health that cannot fall within conventional subjects, was introduced in elementary, junior and senior high schools besides conventional subjects. Schools were supposed to arrange their own curriculums without text-books. However, examination of the state of implementation of the "period of integrated study" revealed that some schools achieved great results while the original purposes and ideas were not fully achieved in other schools, and the overlapping of curriculums among different levels of schools like similar learning activities in elementary and junior high schools, occurred, showing that it is necessary to clarify the purpose of the "period of integrated study" and study about what abilities to be fostered in pupils and how to indicate learning activities.

Schools and teachers themselves must consider curriculum arrangement for the period of integrated study, create the contents, and manage the curriculums. Absence of such experiences in the field of education for the last fifty years or so may be one of the big reasons for the failure of progress. Although there were some practices with results, many practices were simply "looking thing over for study," and mathematical learning contents could not be incorporated.

Therefore, the Central Education Council wrote in the "Summary of Discussion" of November 2007, mainly the following repentance about the current course of study which had promoted the cram-free education.

· Integrated study was not satisfactorily understood by schools.

· The numbers of classes for compulsory subjects decreased.

In the new course of study announced in March 2008, the enrichment of language skills, development and science and mathematics education was proposed, and classes for integrated study decreased while classes for main subjects increased. This course of study is to be implemented from 2011.

3.1.2 Allowance of introduction of advanced contents in courses of study
Introduction of the period of integrated study in 2002 resulted in the reduction of about 30% of the contents of learning of conventional mathematical contents.   Some education-related people supported period of integrated study while other education-related people, parents, teachers and cram schools criticized the reduction of contents of learning of conventional subjects right after the revision for fear of the decline in academic ability. Therefore, the MEXT partially revised the current course of study in 2003, clarifying that the current course of study is the "minimum standard" and that schools are allowed to teach more advanced contents freely.[3]   Thus, schools have become able to recover at least the reduced contents, but very few are doing so. Limitation in the number of hours is one of the reasons, but another reason is teachers' resistance to *teaching contents* beyond textbooks. Many teachers feel reluctant to teach even the contents written in the past textbooks, once they disappeared from the textbooks. Further, since high school entrance examination is limited within the range of the course of study up to junior high school, teachers of junior high schools are busy teaching for the purpose of preparation for examinations rather than spending time to covering contents beyond the range of textbooks.

3.1.3 Study for curriculum development for the enrichment of technology, science and mathematics education   While setting the target on "the cram-free education", the *MEXT* designated 26 high schools all over Japan as *Super Science High Schools* and started priority education for high school students in science and mathematics in 2002.[4]   The budget started with about 700 million yen but it reached about 1.5 billion yen in 2008 and 102 high schools participated nation-wide. This project aims at the promotion of development of advanced science and mathematics education

The MEXT lists the following items as its targets to address.

- Development of curriculums focusing on science and mathematics for the unified junior and senior high school education system by the curriculum arrangement independent of the course of study.
- Implementation of advanced science and mathematics education in cooperation with universities allowing students to attend university classes or university teachers or researchers to give lessons at schools, for example.
- Development of teaching methods and materials for enhancing logical thinking ability, creativity, and originality to even higher levels

One of the characteristics of this project is that an attempt for "Education for Excellence," namely cultivation of human resources with high logicality, originality and creativity, has started, apart from the idea of "Education for All" that Japan has kept consistently after World War II.

In the case of mathematics education, cooperation between high schools and universities often takes place in the following forms.

(1) Mathematicians of university teach university mathematics at high schools. In this case, mathematicians of university talk by turns about their own field of research as topics, or a series of lectures on one topic is given.

(2) High school students continuously attend the mathematics class of a university.

(3) A team of a high school teachers and a university researcher gives lessons at a high school.

The most frequently used approach is the case (1), and the case (3) in which high school teachers proactively take part in teaching and developing curriculums is rare. There is a school, however, where high school teachers are going to take part in lessons proactively in the future supported by university researchers through continued study.

The student advancement rate to high schools is over 96% today, and both students and school characteristics have diversified. Under such a condition, many schools are creating their own arrangement of the contents of the course of study to fit their characteristics.

Since, the development of teaching materials expanding beyond the boundaries of the course of study, such as SSH, has just started, however, many high schools rely on university researchers to give lectures at them. In some cases, high school teachers teach their students using university textbooks as they are. Elite students are reported to have accomplished certain results even when they learned from university mathematicians using the first year university materials as they are. In the future, efforts to provide high-level students with suitable high-level education are desired.

In U. S., President Bush proposed, in his State of the Union address in 2006 the increase of 70,000 teachers to enhance science and mathematics education, however, some people say that there are no concrete measures. [5]   So far, three movements in Japan for realizing more flexible curriculums were introduced.

Development of teaching materials with contents beyond the course of study was found to be still difficult for schools. In fact, this is not the problem of school teachers alone. Many institutes and researches involved in curriculum arrangement share the same problem. Since government-designed textbooks were first prepared in 1901 in the history of Japanese mathematics education, whenever new contents were introduced, textbooks were prepared based on the contents imported from Europe and the US that were completed to some extent. This applied to the early stage of algebra and geometry, the time of introduction of function education, and the time of introduction of modern mathematics. To put it strongly, even today we scarcely have any methodology of creating new teaching contents as Okamori has pointed out [6].

Such a methodology will become definitely important for us when we think of future mathematics education. It will be able to found in the methodology what roles mathematicians, mathematics education researchers, school teachers and others can play.

3.2 The introduction of new mathematics into education

In this section, we explain a viewpoint in order to introduce a new mathematics into the education. In the age of the "modernization of mathematics movement", "new mathematics" were tried to be introduced to schools. A word "new mathematics" in this chapter is not the same as them. The meaning here may be a new "new mathematics" in 1990's. However how it is introduced should be considered based on the "modernization of mathematics movement" in 1960-1970's.

3.2.1 Curriculum arrangement looking at introduction of new mathematics   The problems we face in the drastically changing modern society are large and have many factors entangled complicatedly with each other. Advanced sciences are progressing rapidly by finding new facts and replacing old facts with them. These are problems that cannot be solved easily with fixed knowledge and methods of solution that we have had.

In mathematics, too, new fields have rapidly progressed in the latter half of 20th century accompanied by the development of the computer. For example, graph theory, one of the fields of discrete mathematics, has rapidly progressed accompanied by the development of the computer. Further, knot theory has been studied vigorously since

1980's and is applied in various fields today.

When we think of the current situation like this, we find that the power of challenging to solve unknown problems is important in surviving the future society. However, the school mathematics that Japanese students learn today is fixed, already systematized, and completed as one of the fields of mathematics. The lessons are focused on understanding of knowledge, memorization of solution patterns, and calculus. Thus, students have no doubt about what they learn. Many students think that there is a definite answer to any problem of mathematics.

Having students touch fresh studies as mentioned above and their processes of creation may work as one of the measures to overcome such situation of mathematics education. Therefore, when we think of arranging a curriculum, we have to consider 'what', 'when' and 'how' from the wide range including new mathematics, unrestricted within the fields currently treated in the curriculum. As the first step of curriculum arrangement, it is important to consider how to make it possible to introduce new mathematics. For example, in US, the introduction of discrete mathematics into education was argued in the year book of NCTM in 1991. [7]

3.2.2 What we learn from modernization of mathematics education   On the other hand, when we think of the introduction of new mathematics in the course of study, we can look back at the *"modernization of mathematics movement"* in 1960's.

The modern math movement of education started in US spread out in the world within about ten years in 1960's. In Japan, the course of study started to revise in 1968 taking reference to SMSG, textbooks of US, and SMP, textbooks of UK. Following points were emphasized in curriculums for modernization.

(1)Realization of consistent arithmetic and mathematics education through the
   introduction of concepts and ideas of modern mathematics, namely set, function,
   mathematical structure, etc.

(2) Knowing the merit of mathematical view and the way of thinking

(3)Strengthening the logical thinking power rather than the quick and correct calculation

It is true that modern mathematics movement of education led by mathematicians increased dropouts.

Many criticisms appeared including the following claim by Kodaira. [8]

> *"Advocates of modern mathematics education may say that mathematics education should be started from set since the base of modern mathematics is the set theory. ....Logical base does not necessarily work as the base for learning. ....Its basic idea is too difficult for elementary school and junior high school students to understand."*

As a result, the course of study which aimed at the modernization of mathematics education was forced to change direction to the "cram-free education" in 1977.

Then what can we learn from the modern mathematics movement of education? It is true that one of the points of repentance was that the modern mathematics movement of education was so hasty that modern mathematics was taught to students without modification. Then how should students learn?   In order to find the answer, we can refer to Freudenthal's argument that is emphasizing to observe the learning processes.[9]

> *"How should children learn?, in particular mathematics, which I immediately change into "How do people learn?", which is the proper question, and the way to answer it would be:*
>
> *By observing learning processes, analyzing them and reporting paradigms – learning processes within the total educational system, learning processes of pupils, groups, classes, teachers, school teams, councilors, teacher students, teacher trainers, and of the observers himself.*
>
> *Observing involves analyzing, by which I don't mean averaging or applying other statistical procedures nor fitting the observational data into preconceived patterns of developmental psychology, Grasping "how people do learn" would be a first step towards solving the everyday problems of practitioners. Grasping "how to teach learning", and "building a learning theory", should be based on evidence, rather than on preconceives ideas."*

Thus, the question "How should students learn?" should be found in "mathematics education as a field science".

3.3 An approach to the development of teaching knot theory

When we develop materials for an education of new mathematics, we should not provide students with mathematics as it is, but we should modify it to "mathematics for education" taking students' understanding into consideration, without sticking too much to rigor and a system. Its educational meaning should not be that the mathematics is "simply new," but it should be fully discussed and clarified. Which concept of the mathematics is meaningful to students and how to introduce it to fulfill its meaning should be discussed. Through such a process we can make up "mathematics for education" which may be a little different from the original way of expression and arrangement.

In order to tackle such work, it is important to plan a project with a team work of professional mathematicians who understand education, education practitioners who understand mathematics, and mathematics education researchers as professional epistemologists concerning mathematics. Each member should take part in the project with equal status respecting and understanding the others' professions.

The professional practitioners of education, namely teachers, are not only high school teachers, but also include elementary and junior high school teachers, and they need to establish a system by seeing the degree of the understanding of mathematics from lower grade students to higher grade students.

In 2004, we planned a project like mentioned above, and have discussed the introduction of a field of new mathematics, knot theory, into education. This project was started as a part of research of A. Kawauchi, who was the leader of "Constitution of wide-angle mathematical basis focused on knots", one of the 21st Century Center of Excellence Programs, from April 2003 through March 2008. The project members were Kawauchi, 5 researchers of mathematics and mathematics education in university, 3 high school teachers, 2 junior high school teachers, 3 elementary school teachers, and 2 graduate students. All the members majored in mathematics either at Osaka Kyoiku University or at Osaka City University. The teachers who graduated from Osaka Kyoiku University also have licenses of elementary, junior and senior high school teachers of mathematics. In the first three years, an experimental teaching was carried in elementary, junior and senior high schools attached to Osaka Kyoiku University and private high schools. Further experimental teachings were done in Osaka City University and a public high school based on a collaborative teaching program between University and Senior High School as a *Super Science High School* Program in 2007. And in 2009, 2 former school principals and 5 junior high school teachers joined our project. We started to teach knot theory in public junior high schools based on the results of experimental teachings of the elementary school and junior high school attached to Osaka Kyoiku University.

We started tackling the problem in the following processes while having meetings once a month. In each process, all the members studies and discussed from their own standpoints. I think that we can think of the meaning of materials of each step aiming at the systematization of a series of *teaching contents* by the cooperation of members from elementary schools and universities.

(1) An expert of knot theory gives a lecture on the basics of "knot theory" to other university researchers and teachers. Here, the history of the development of "knot theory," its position in the present fields of mathematics, and its meaning in the

fields of mathematics such as examples of applications in various sciences are introduced.[10]   Members except the experts of knot theory experience "learning of knot theory" and also study by themselves using primers. Then they discuss educational meanings of knot theory based on the "observations of their own learning processes".

(2) Experts of mathematics education take leadership in proposing the educational meaning of knot theory and contents that can be used as materials for elementary, junior high and high schools. All the members discuss from their own standpoints under mutual respect as equals.   Mathematicians provide mathematical topics related to the proposals, while school teachers further propose substantiation of materials.

(3) In order to confirm educational meanings and utility of the materials, researchers of mathematics education propose surveys of students' understanding as necessary.

(4) School teachers carry out surveys of understanding. Researchers of mathematics education, teachers, and graduate students analyze the results.

(5) Based on the results of the survey, researchers of mathematics education and school teachers select concrete materials and discuss and propose arrangement.

(6) Researchers of mathematics education and teachers design the classes, propose necessary tools and work sheets, and graduate students prepare them.

(7) Teachers carry out educational experiments. University members visit the class and take records as necessary.

(8) All the members consider the results of class analysis and educational experiments.

Elementary, junior and senior high school teachers hold meetings at the school levels with researchers of mathematics education to review surveys of understanding, materials preparation, and educational experiments.   Based on the results obtained through the processes mentioned above, we find new challenges such as students' understanding, materials, teaching methods, mathematical problems of knot theory itself. Then we repeat the processes and accumulate teaching materials.

3.4 Conclusion

In the history of mathematical education of Japan, there was almost no experience of developing contents of education for ourselves from the beginning. Therefore, schools and national institutes for selection of educational contents presently have not found the methodology yet.   We described in this paper, a methodology of development of educational contents of new mathematics. It can be said a methodology of field science.

In this methodology, researchers of mathematics education and mathematicians cooperate respecting and trying to understand researches of others' professional fields and create materials for education. Today, in Japan, no other organization is found that studies with such methodology.

It is found the following facts so far in the project. It is not so easy to create mathematics for education.

・ We depend on researchers of mathematics education and an expert of knot theory for the large part. We need to grasp firmly children's understanding on mathematical ideas in the field of school rooms and at the same time to understand essential aspects of the mathematics. Researchers of mathematics education propose educational meanings of teaching knot theory based on the pupils' cognitions. The expert of knot theory who plays leading roles on a theory of mathematics can tell all the members about their "creative activities" on how the theory was created, including experiences of surprises and difficulties. They are quite different from indirect understanding by reading technical books of completed theories. It is important for researchers of mathematics education and teachers to study the mathematics actually even up to an introductory level. From doing this, ideas for educational mathematics will be produced. This is necessary even for elementary school teachers so that they are required to have ability of understanding mathematics to some extent.

・ School teachers played a role of making the teaching plan come true.

Such a method of materials development requires repetition of proposals and experiments to improve materials. Further, materials of next stages must be prepared to fit the growth of children. It is needless to say that it is important to develop good materials with this method, and at the same time, the experiences of people who are in the educational field researching the development of materials according to the methodology are also important.

References

[1] Course of Study for Elementary and Junior High Schools (in Japanese), MEXT. 1998.

[2] Course of Study for High School (in Japanese), MEXT. 1998.

[3] Partial Amendment of the Course of Study 2003 (in Japanese), MEXT, *at* http://www.mext.go.jp/english/shotou/index.htm.

[4] Document on SSH (in Japanese), MEXT, *at* http://www.mext.go.jp/b_menu/houdou/20/04/08040905/004.htm.

[5] A Trend of Education in the World (2006) (in Japanese), MEXT. 2007.

[6] H. Okamori(ed.), Research and Practices of Mathematics Education (in Japanese), DAI-ICHI HOKI co. ltd., 1983.

[7] Discrete Mathematics across the Curriculum K-12, NCTM, 1991.

[8] K. Kodaira, A Criticism of the "New Math" (in Japanese), Science, 10, Iwanami Shoten, Publishers, 1968.

[9] H. Freudenthal, Major problems of mathematics education, Educational studies in mathematics, 12, 1981.

[10] A. Kawauchi, Knot Theory (in Japanese), Springer-Verlag, 1990.

[11] T. Yanagimoto, Y. Seo, K. Iwase, "A Study on Making Knot Theory into a Curriculum", Proceeding of the 5th East Asia Regional Conference on Mathematics Education (EARCOM5), Vol.2, 637-644, 2010.

4. Education Practice in Elementary School

In this chapter, we introduce an experimental teaching of knot theory in an elementary school. In 4.1, a significance of teaching knot theory in the elementary school is considered. In 4.2, the result of an investigation on drawing a knot and a cube is reported. In 4.3, we state the teaching contents of knot theory done for pupils in an elementary school.

4. 1 A significance of teaching mathematical knot in elementary school

A mathematical knot has a possibility to be an effective material in mathematics education in schools. In particular, it is considered to be effective in training pupils' spatial visualization in the elementary school. A spatial visualization is essential in analyzing knots since knots are 3-dimensional figures in themselves. Here, it is explained by the following two examples.

4.1.1 Looking at a knot from different viewpoints
We can start with imagining a knot as a 3-dimensional figure from its diagram. A diagram of a knot indicates the positional relationship of the parts of a knot divided by the crossing points. It is not so easy to imagine how the crossing points look like from other directions. For example, let's look at a knot from two directions as it is illustrated in Fig.4.1. When Fig.4.2 indicates the diagram seen from the viewpoint A, the diagram of the knot seen from the viewpoint B, the opposite side of A, becomes like the diagram in Fig.4.3.

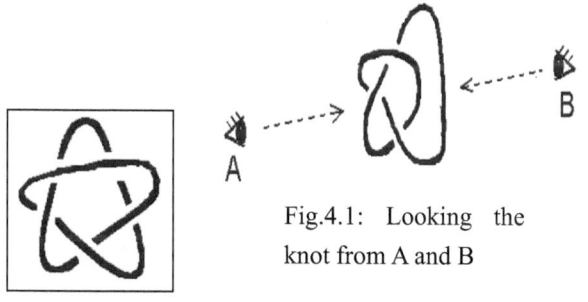

Fig.4.1: Looking the knot from A and B

Fig.4.2: The diagram looking from A

Fig.4.3: The diagram looking from B

When the figure seen from the viewpoint of A is called "the original knot diagram" seen from the front, the diagram looking from the viewpoint of B can be called "the knot diagram looking from the opposite side of the original one".

The knot diagram looking from the opposite side is the one seen from our viewpoint by a 180 degree rotation on a vertical axis. It can be also said that the diagrams of the knot are obtained from each other by the opposite 180 degree rotation of the knot.

We can show another example of deformations by moving parts of a knot. The "figure-eight knot" (Fig.4.4-a) is a knot with the crossing number four, the smallest except the trivial knot and the trefoil knot. If we look it from the opposite side, like in Fig.4.4-b, we find that it is the same diagram as the original figure-eight knot. Furthermore, it is found that the mirror image of the original figure-eight knot (Fig.4.4-c) is "the same knot" as the original one.

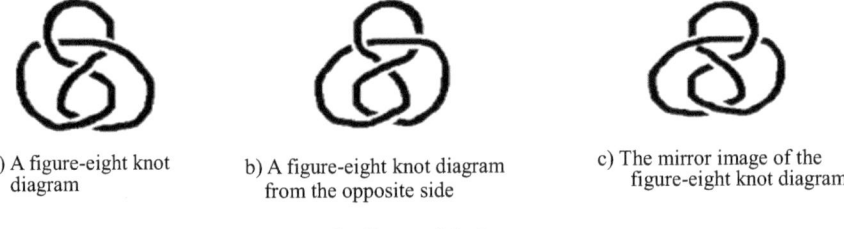

a) A figure-eight knot   b) A figure-eight knot diagram   c) The mirror image of the
   diagram   from the opposite side   figure-eight knot diagram

Fig.4.4: The figure-eight knot

The following sequence of moves indicates how to obtain the original figure-eight knot from the mirror image by using rotations only.

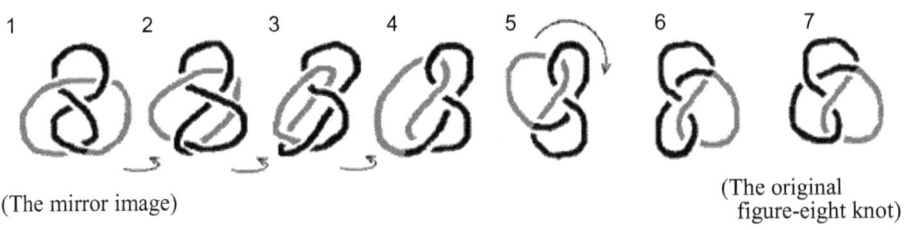

(The mirror image)   (The original
                                                                figure-eight knot)

Fig.4.5: The moves from the mirror image to the original figure-eight knot (1)

On the other hand, we can also find that the mirror image of the original figure-eight knot is "the same knot" as the original one by using deformations of parts of the knot and rotations as follows:

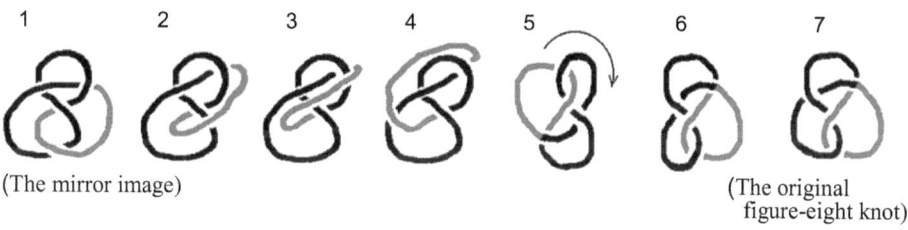

(The mirror image)   (The original
                                                                         figure-eight knot)

Fig.4.6: The moves from the mirror image to the original figure-eight knot (2)

From the above, we find that through the activity thinking of knots, we can frequently move figures in the three-dimensional space and observe them by changing viewpoints, which mean that we are observing 3-dimensional objects while handling them dynamically.   This kind of 3-dimensional approach was hardly seen in the traditional mathematics education. In the current when knot figures in the space were studied, they were almost always handled statically. In observing the faces, edges and vertexes of polyhedra, some models are sometimes used, but even such a chance does not come so often.

In the study of a space vector at senior high school, a static figure in the space is analyzed with the concept of a vector.   A body of rotation grasps a figure in the space dynamically as a body of rotation of a plane figure, which can be said that it grasps 3 dimensions statically as the dynamic locus in 3 dimensions.   On the other hand, a series of deformations of a knot as mentioned in Figs.4.5-4.6 can be said to be exactly the dynamic observation of a knot with the passage of time considered, so that we do hope that if learners have gotten used to this kind of dynamic handling of figures in the space at an earlier stage, they could get much better spatial visualizations.

In moving and observing a 3-dimensional figure, there is a danger of moving the parts of them vaguely without awareness. Thus, expressing the process of moving them by a diagram leads to the awareness of the parts of the 3-dimensional figures and to getting an image of a next move from the diagram.   Expressing the process of observing and thinking by a figure helps learners themselves deepen their thinking, and communications with others lead them to heighten their thinking as well.   Expressing by a figure can exactly be considered to be an important means of thinking.

It is not so easy for pupils in elementary schools to draw a sketch of a regular parallelepiped or a cube by themselves in fast.   When we draw a sketch of it, it is necessary to express the angles and the ratios of the lengths among the edges. In case a balance of these quantities breaks down, such a sketch is no longer regarded as a regular parallelepiped or as a cube.

On the contrary, unlike a solid handled in the traditional figure teaching, a knot is marked by the fact that it is easier for pupils to express it by a diagram. Knots in 3-dimensions are expressed as a gathering of several arcs in 2-dimensions. As for crossing points where two arcs overlap, as long as the upper arc and the lower arc are expressed without mistakes, it does not matter so much to draw the arcs by taking a wrong length and a wrong turn. The upper and lower arcs on the crossing points alone express the 3-dimensional.

Accordingly, we think that knots can become figures in the 3-dimensional space easier to express to pupils.

Based on the points mentioned above, we came to the following hypotheses.

(1) Knots could be helpful in educating pupils' spatial visualization through handling figure dynamically.

(2) We wonder whether a knot is easy-to-draw in comparison with a familiar solid such as a cube or a regular parallelepiped.

If so, is it possible to improve pupils' spatial visualization through knot theory at an earlier stage than when they have come to be able to draw a cube?

4.2 Investigation of the pupils' drawing

We made an investigation on the 192 pupils in total from the 1st grade to the 6th grade in an elementary school.

4.2.1 Questions for investigation The purpose of this investigation is to find which one of the cube or the trefoil knot is easier for pupils to draw. We asked them the following questions.

Q1: Let's draw a cube, placed on a desk, whose edge is 3cm, just as it is. (Photo 4.1)

Q2: Let's draw a string that can be seen in a box by paying a careful attention to "the overlapping of the string". (Photo 4.2)

Photo 4.1:
How a pupil is drawing a cube

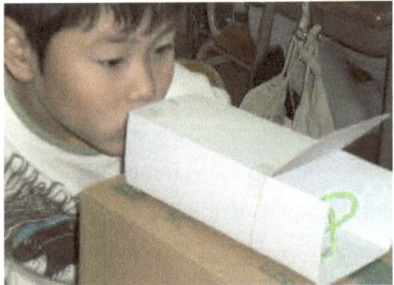

Photo 4.2:
How a pupil is taking a look at
a string (the trefoil knot) in a box

When we get pupils draw a knot, it is important to keep their viewpoints from moving. Otherwise, the knot diagram is changed. Therefore we set a knot in a box and get pupils look into the box. Furthermore, we made a part of the top of the box open to give a 3-demensional appearance to the knot by a light.

4.2.2 The result of the investigation Through the investigation above, the following points are found. On Q1, we classified the pupils' drawings into the 5 types as shown in the following data.

Type A: Each pair of edges is parallel and each angle is not a right angle.
Type B: Each pair of edges is parallel but some angles are right angle.
Type C: Some pairs of edges are not parallel.
Type D: Adjacent sides are drawn in a same straight line.
Type E: The connection between faces is wrong.

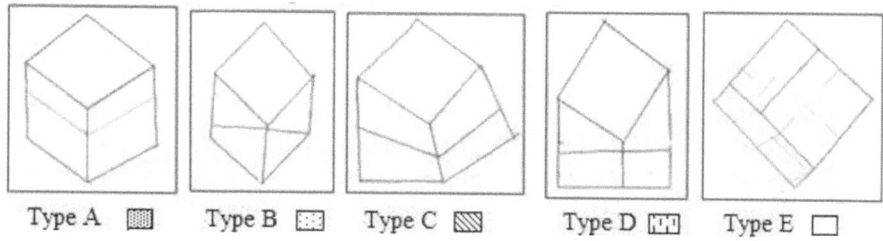

Type A ▦   Type B ▦   Type C ▧   Type D ▦   Type E ☐

Fig.4.7: The types of the pupils' drawings of the cube

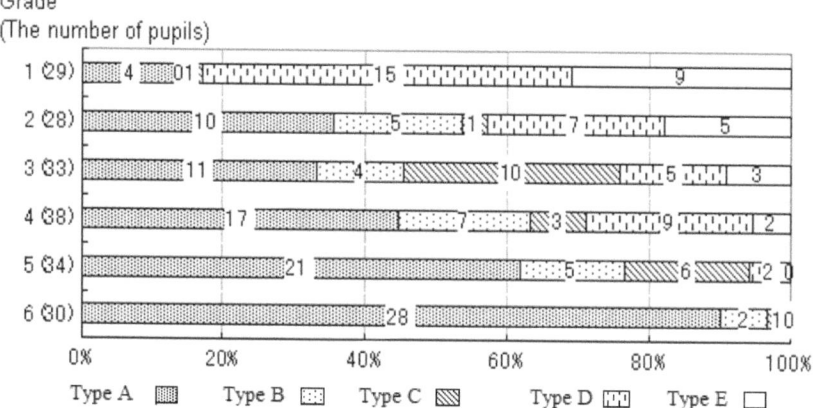

Table 4.1: Drawing the cube

44

On Q2, we classified the pupils' drawings into the 4 types as shown in the following data.

      Type A: The drawing is almost correct

      Type B: Though the drawing is almost correct, only an overlapping point is upside down.

      Type C: All overlapping points are not drawn two-tiered.

      Type D: Others.

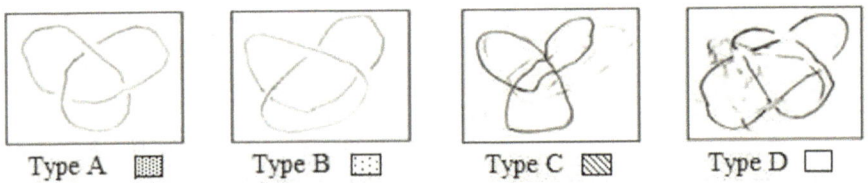

Type A ▦   Type B ▦   Type C ▧   Type D ☐

Fig.4.8: The types of the pupils' drawings of the knot

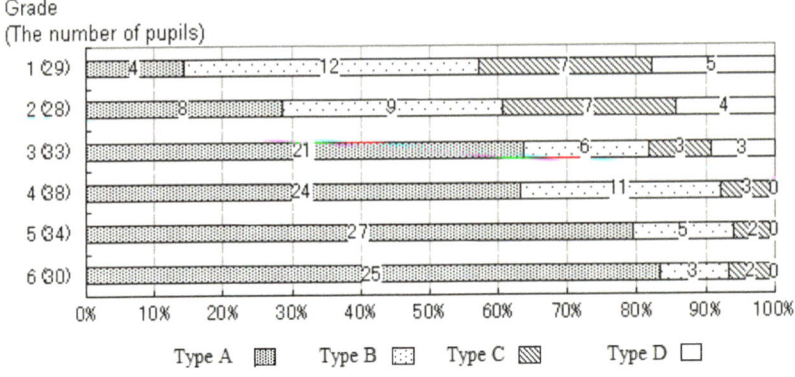

Table 4.2: Drawing the trefoil knot

We found out of this investigation that it was easier for the pupils at middle grades in elementary school levels to draw the trefoil knot than the cube.

4.3 Education practice on knot as a 3-dimensional figure

In this practice here, we experimented with the 4th grade pupils in an elementary school and created teaching materials aiming to improve their spatial visualization through concrete activities with knots.

1) Look at an actual knot and draw the *knot diagram*:   When a knot is to be expressed in 2 dimensions, it is related to a sketch made under an existing mathematical idea. However, in the case of an actual knot, the situation of the string placed before and behind shown as a stereogram can be distinctively expressed only by the up-and–down movements of the string.

2) Make a knot and imagine it out of the figure of the knot drawn on paper: Contrary to 1), this leads them to think of transforming a 2-dimensional figure into a 3-dimensional figure.

3) Move a part of knot, imagine the transformations and express in the diagrams: There are "sliding", "twisting", "rotating" and "shrinking" among the concrete movements of a part of the knot.

4) Imagine figures seen by changing viewpoints on a given knot and express them in diagrams: Imagine a knot seen from the opposite side and a 180 degree rotation of a knot along an axis..

5) Imagine the mirror image of a knot and express it in a diagram: This leads to thinking of the plane symmetry of figures in the space.   As for "a figure-eight knot", in particular, the original knot and the mirror image become the same knot. This can be confirmed through using an actual string and expressing them as diagrams.

In order to create the teaching contents of experiments based on the activities mentioned above, we set up a series of the teaching contents as follows;

1) Picking up easy knots with 3 crossing points to observe the difference between "a trivial knot" and "non-trivial knots".

2) Picking up trivial knots that have 2 or 3 crossing points and express in diagrams how the knots are being untied.

3) Finding "the same knot" as a trefoil knot and see how they can make "the same knot" by moving a part of the knot and express the deformation process as diagrams.

4) As for knots with up to 5 crossing points, imagine "the knot seen from the opposite side" and express them in diagrams.

5) As for knots with up to 5 crossing points, imagine their mirror images and express them in diagrams.

6) To confirm that the figure-eight knot is the same as its mirror image, make observations through moving a part in a diagram and through changing a viewpoint.

Experimental teaching was conducted on a class of the 4th grade 38 pupils in the elementary school attached to Osaka Kyoiku University for 5 school hours.

4.3.1 Lessons and the pupils' work   The following activities are carried out in the lessons.

Lesson 1: *Untie a knot* (2 school hours)

1) Look at actual knots and think which one can be untied (Fig 4.9).   Also, think of which parts of the knot should be moved for untying.

Fig.4.9

2) Draw the following knots like an animation-type comic and untie them (Fig. 4.10).

Fig.4.10

The pupils' reaction:

Pupils traced the first diagram and revised a part of the knot where they deformed. Then they traced the second diagram and revised a part of it. They repeated this procedure paying attention to the overlapping point.   Fig.4.11 is an example of a pupil's work.

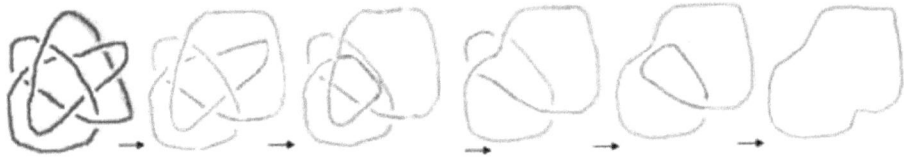

Fig.4.11

Lesson 2: Think of the knots seen from the opposite side (1 school hour)

1) As for the strings in the right, imagine how the look is seen from the opposite side, and draw them.

Fig.4.12

2) As for the knots below, imagine how they are seen from the opposite side, and draw

them.

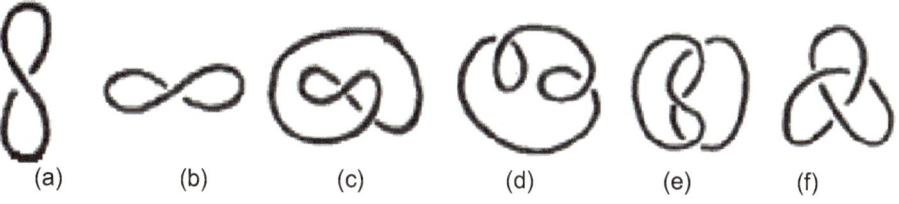

(a)   (b)   (c)   (d)   (e)   (f)

Fig.4.13

The pupils' reaction: The pupils imagined the diagram and drew it first. Then they confirmed the diagram by using a soft wire (Photo 4.3). They could imagine the knot seen from the opposite side through these activities

Photo 4.3

Lesson 3: Find the *same knots* (1 school hour)

We introduced the two "*trefoil knots*" below and presented that these were not the same knot.

The trefoil knot *A*   The trefoil knot *B*

Fig.4.14

Based on this remark, we make them think which of the knots from (a) to (f) can be identified with the trefoil knot *A* or the trefoil knot *B*.

(a)   (b)   (c)   (d)   (e)   (f)

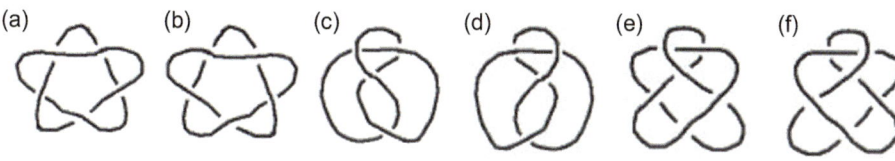

Fig.4.15

For this purpose, we make the pupils think in the following procedure.

(1) First, draw the process of deforming each knot diagram into a trefoil knot diagram.

(2) Deform the trefoil knot diagram into the original figure with a soft wire by using the reverse process of the procedure (1).

The pupils' reaction: The diagram of Fig. 4.16 is a pupil's work. It indicates the knot (f) is the same knot as the trefoil knot A.

Pupils imagined the diagram and drew it first. Then they confirmed the diagram by using a soft wire (Photo 4.3).   They confirmed that the two knots are the same each other by deforming the trefoil knot to the knot from (a) to (f) (Photo 4.4 and 4.5).

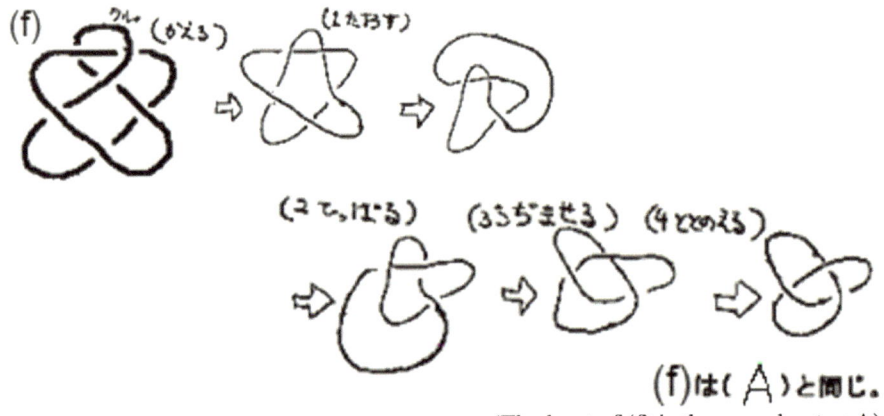

(The knot of (f) is the same knot as A)

Fig.4.16

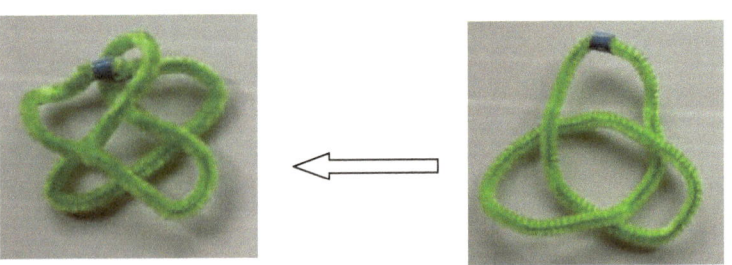

Photo 4.4: The knot of (f)   Photo 4.5: The trefoil knot *A*

Lesson 4: Think of the *mirror image* of a knot (1 school hour)

As for the knots of (a) to (g) below, think, imagine and draw how they look when the mirror image of each knot diagram is placed in the right side of the dotted line (Fig.4.17).

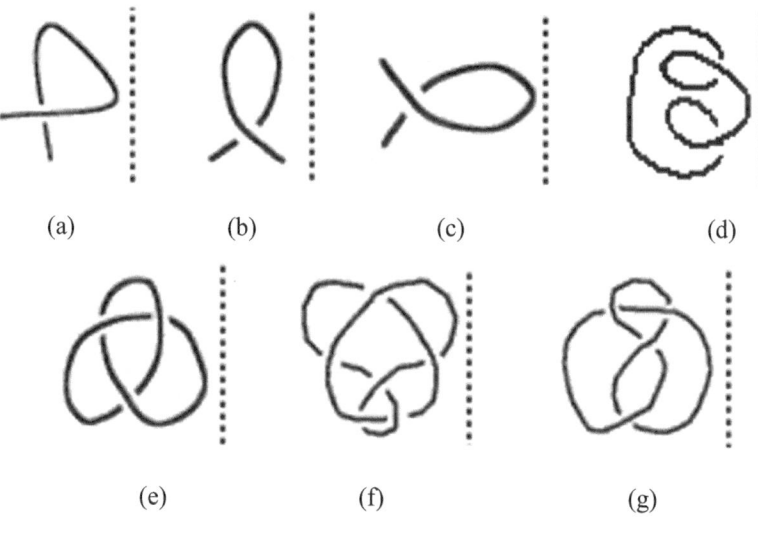

(a)   (b)   (c)   (d)

(e)   (f)   (g)

Fig 4.17

The pupils' reaction: Pupils confirmed their diagrams by using a plastic plate (Photo 4.6).   It is easier for them to imagine the mirror image of the knot than to see the knot from the opposite side.

Photo 4.6: Using the plastic plate in order to confirm.

4.4 Education practice on reducing crossing number of a knot

Based on the experimental practice done above, we designed a teaching content related to *reducing the crossing number* of a knot diagram. We tried to teach two 5th grade pupils as an experimental teaching.

4.4.1 Lessons and the pupils' work   The following activities are carried out in the lessons.

Lesson 1: Assigning the two arcs around each double point of each diagram to an upper arc or a lower arc in order to obtain an untied knot.

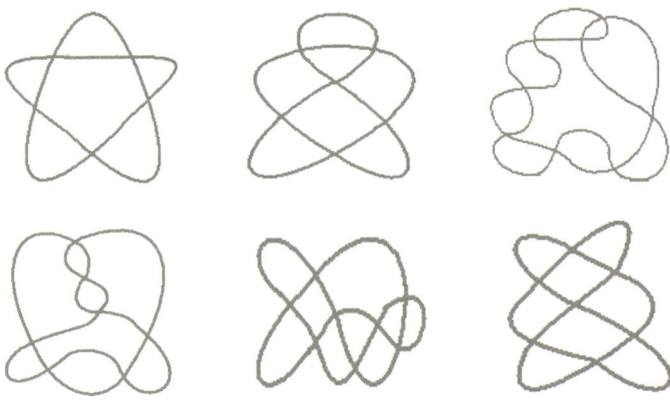

Fig.4.18

The pupils' reaction: It was easy for them to make knots with crossings. They could imagine untying their knots as follows:

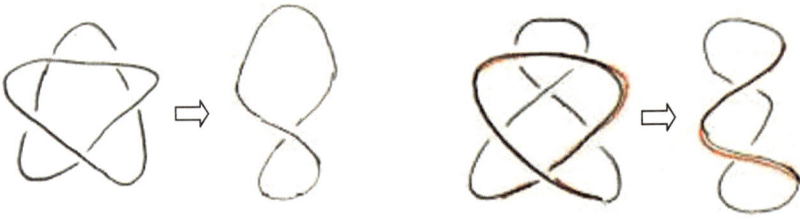

Fig.4.19

Lesson 2: Untie each knot by using a "magic".   -part 1-

We first explained the "magic" which means exchanging the upper and lower arcs around the crossing point.

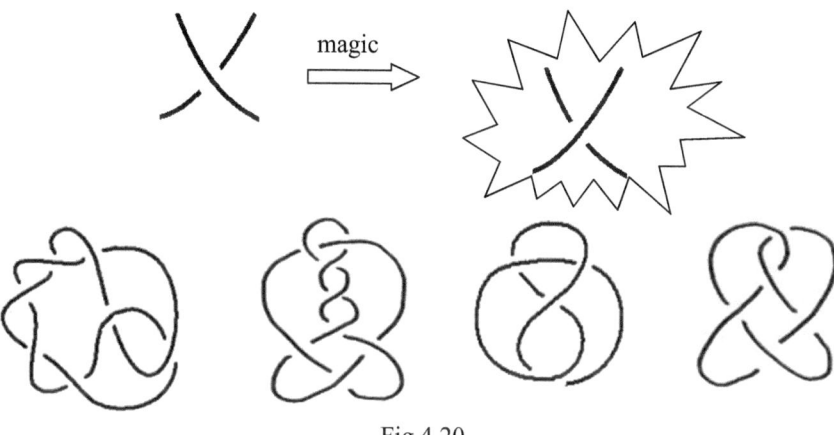

Fig.4.20

The pupils' reaction: Pupils first focused on the twisting part of the knot to use the Reidemeister move I, in order to untie the knot (Fig.4.21). The number in [   ] of the Fig.4.21 indicates the exchanging number of the crossing points.

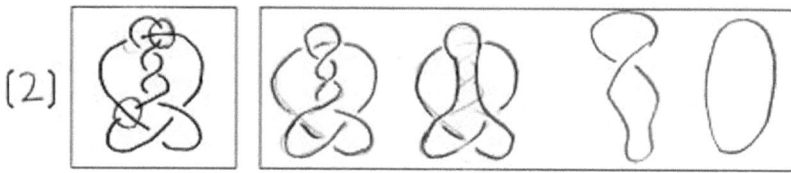

Fig.4.21

We made them think to make the exchanging number of the crossing points fewer (Fig.4.22).

Fig.4.22

Lesson 3: Untie each knot by using the "magic". -part 2-

We get them think the same point as in Lesson 2. However, the number of the crossing points of the knot diagrams in Lesson 3 is more than in Lesson 2. These knots seem to be difficult to find the exchanging points for untying (Fig.4.23).

Fig.4.23

The pupils' reaction: A pupil first focused on the 2 crossing points and tried to untie the knot (Fig.4.24). Then he noticed that the resulting knot is a trefoil knot. Thus, he challenged again by focusing 3 crossing points and succeeded to untying it (Fig.4.25).

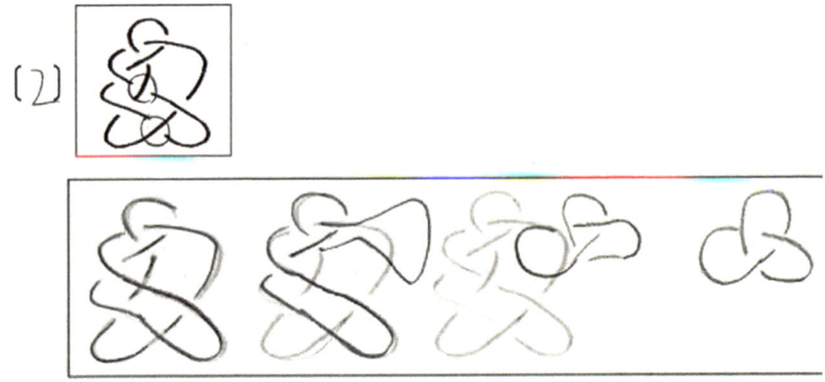

[2]

Fig.4.24

[3]

Fig.4.25

Lesson 4: Consider how to use the magic effectively.

We introduce a way how a knot is untied easily.   In this way, we always go through along the under arc when we go along the string of a knot diagram and meet a crossing point. Here, it is not needed to exchange the upper and lower arcs around a crossing point.

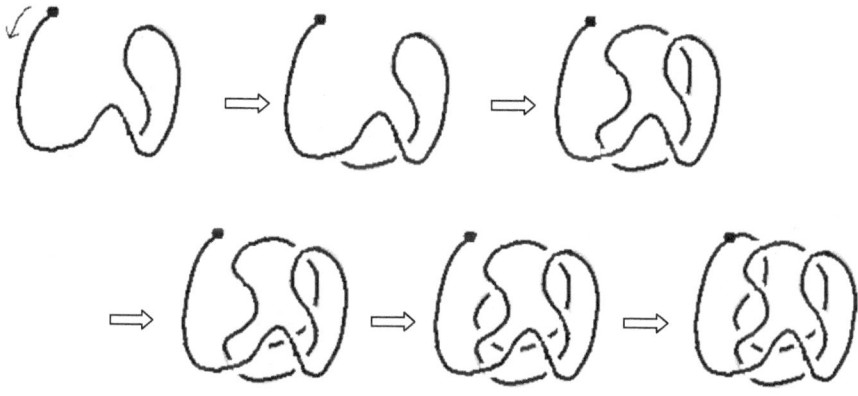

Fig.4.26

The pupils' reaction: They think why this drawing way always makes the knot untied.   We showed two actual knots in each box with the same projection image so that one is trivial knot with a diagram drawn as in the way above and the other cannot be untied.   The top side of the box is a transparent plastic plate and a side can be taken off.

Photo 4.7:
A tied knot in a box

Photo 4.8:
An untied knot in a box which is taken a side off.

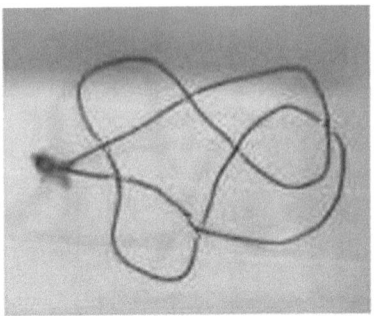

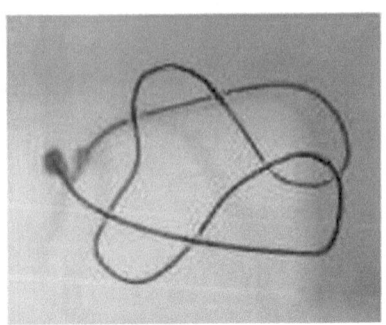

Photo 4.9:
A tied knot in a box viewed
from directly above

Photo 4.10:
An untied knot in a box
viewed from directly above

At first, the pupils traced the knots on a clear sheet looking from the upper side (Fig.4.27 and 4.28). Then they guess which can be untied. The next, they investigated the knot in the box from the side and were surprised that the trivial knot seems very simple when they looked it end-on. (Photo 4.11 and 4.12)

Fig.4.27

Fig.4.28

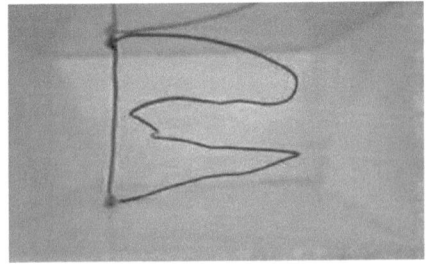

Photo 4.11

Photo 4.12

4.5 Conclusion

Through our experimental practice at the elementary school, we generally found the following points.

(1) By looking at the knot diagrams and through the activity of actually making knots by a soft wire, the pupils become able to grasp the knot diagram drawn on paper 3-dimensionally and dynamically.

(2) Handling knots led them to be able to deal with rotational and plane symmetric moves.

(3) They became aware of the symmetry of the mirror image, and for some knots with crossing points, they learned to be able to imagine the mirror images from their figures and express them in diagrams.

(4) As for thinking of "the knots seen from the opposite side", they could imagine them soon after studied while using actual strings, so that we cannot necessarily say that this is a difficult content to the 4th grade pupils in elementary schools. However, actually, half of them could not answer our research questions given in 2 and 3 weeks later. This led us to think that "the knots seen from the opposite side" require not a just 1 school-hour lesson but hours of repeated studies until it can come to stay with pupils.

According to these results, we think it is possible to make pupils imagine the moves of knots in the space from the elementary school, which enable us to improve pupils' spatial visualization.

References

[1] A. Kawauchi et al, Knot Theory (in Japanese), Springer Verlag, Tokyo, 1990. (English expanded version: A Survey of Knot Theory, Birkhäuser Verlag, 1996.)

[2] A. Kawauchi and T. Yanagimoto et al, An Approach to Teaching Knot Theory in School Mathematics for Pupils and Students (in Japanese), Project of Teaching Knot Theory in School Mathematics, Research Report as Educational Action in 21st Century COE Program "Constitution of wide-angle mathematical basis focused on Knots(Osaka City University)" Vol.1 2005, Vol.2 2007, Vol.3 2009.

[3] C. C. Adams, The Knot Book, W.H. Freeman and Company, 1994. (Japanese version translated by T. Kanenobu, 1998.)

[4] A. Kawauchi (ed.), Mystery of knot theory (in Japanese), in: Have Fun With Mathematics, No.5, 26-80, 1998, NIPPON HYORONSHA CO., LTD. PUBLISHERS,

[5] A. KAwauchi, Lectures on Knot Theory (in Japanese), KYORITSU SHUPPAN CO., LTD,

2007

[6] M. Wakayama, Working Mathematics in Technology, Iwanami Press., 2008

[7] T. Yanagimoto, Y. Seo, K. Iwase, M. Terada and R. Kaneda, An Approach to Teaching Knot Theory in Schools, Proceeding of the 4th East Asia Regional Conference on Mathematics Education (EARCOM4), 411-416, 2007.

[8] T. Yanagimoto, Y. Seo, K. Iwase, "A Study on Making Knot Theory into a Curriculum", Proceeding of the 5th East Asia Regional Conference on Mathematics Education (EARCOME5), 637-644, 2010.

Acknowledgements

We wish to express our gratitude to Mr. Shimonosono (Seifu Senior High School) and Mr. Takebayashi (Former Tennoji Elementary School Attached to Osaka Kyoiku University) for cooperating in our education practice.

5. Education Practices in Junior High School

In this chapter, we introduce 2 education practices of knot theory in the junior high school. The first one is a teaching in a junior high school attached to Osaka Kyoiku University. The second one is teachings in public junior high schools, which are based on the education practices in elementary and junior high schools attached to Osaka Kyoiku University.

5.1 Education practice in junior high school attached to Osaka Kyoiku University

In this section, we show our practices in Tennoji Junior High School attached to Osaka Kyoiku University. Main teaching contents are the tri-colorability and the linking number. Both of them are known to be invariants of a link.

5.1.1 Tri-colorability and Linking number   At first, these invariants are briefly introduced as *teaching materials*.

Tri-colorability   We say that a diagram of a knot or a link is *tri-colorable* if each of the upper arcs in the diagram is colored by one in the three different colors so that at least two colors are used and, at each *crossing,* either three different colors come together or the same color comes together. The *trefoil knots* are tri-colorable (Fig.5.1.1).

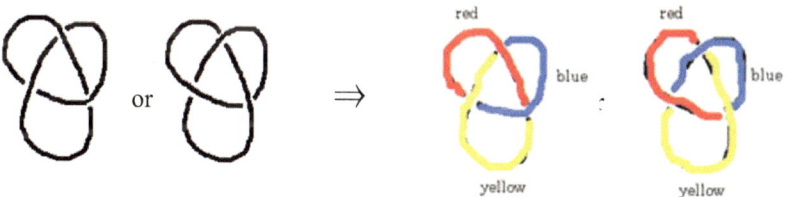

Fig.5.1.1

Even if you deform the trefoil knot into another form, the knot is tri-colorable too. Using the tri-colorability, we can see that some knots are non-trivial knots. Students can make a *conjecture* that tri-colorable knots are non-trivial knots. When a student find a tri-colorable knot, it must be a non-trivial knot. On the other hand, a knot which is not tri-colorable need not be a trivial knot. This condition is logical, and students can find easily non-tri-colorable knots which are known to be non-trivial such as the figure-eight knot. Though it is found that the tri-colorability is unchangeable by

Reidemeister moves (Fig.5.1.2), this education practice is not extended further.

Fig.5.1.2: Reidemeister moves

In 1926, the German mathematician *K. Reidemeiste*r observed that if we have two distinct diagrams of the *same knot*, then we can obtain from one to the other by a series of Reidemeister moves and planar equivalences. Although each of these moves changes the diagram of the knot, it does not change the knot itself.

Linking number Let M and N be two components in a *link* with an orientation on each of them. Then at each *crossing* between the components, one of the pictures in Fig.5.1.3 holds. We count +1 if the *crossing* is of the first type and -1 if the *crossing* is of the second type, and take the sum of these numbers ±1 over all the *crossings* between M and N and then divide this integer by 2. This number is the *"linking number"*.

+1   -1

Fig.5.1.3

5.1.2 Education Practice Purposes of the education practices are as follows:
(1) Let pupils consider mathematics using *strings* by '*trying and erro*r'.
(2) Let pupils go through a stage from a simple problem to a complicated problem.
(3) Let pupils find 'their conjectures' in mathematics.
(4) Let pupils consider 'their conjectures' logically.
(5) Let pupils try to understand the 3-dimensional space.

5.1.3 Teaching plan (1) The main theme of the first teaching plan is the tri-colorability. In the practice here, we experimented with the 3rd grade pupils in the junior high school. The lessons consisted of the following 4 periods:
The 1st period: Let's make a knot.
The 2nd period: Let's untie a knot on a paper.
The 3rd period: Let's find a non-trivial knot.

The 4th period: Let's make a tri-colorable knot.

The 1st period

(1) Let's grasp each other by the hands of 3 persons.   Can you make a circle without taking off your hands?

⇒Let's draw pictures of 3 persons on a paper.

Fig.5.1.4

Photo 5.1.1

⇒Let's draw the pictures simpler.

(a) 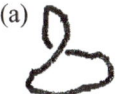   (b) 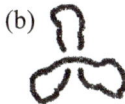   (c)

Fig.5.1.5

In the figures 5.1.5, (c) is impossible to make a circle. We explained the pupils that we call these figures "knots" and we shall think whether or not we can untie them.

(2) Let's find knots that we can untie in the following 4 kinds of knots.

⇒ Let's make these 4 knots by using 4 strings and then find out untied ones among them.

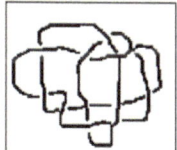   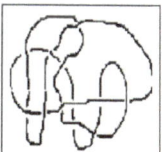

Fig.5.1.6

Photo 5.1.2: Pupils' work

The 2nd period

(1) Let's untie a knot on a paper.

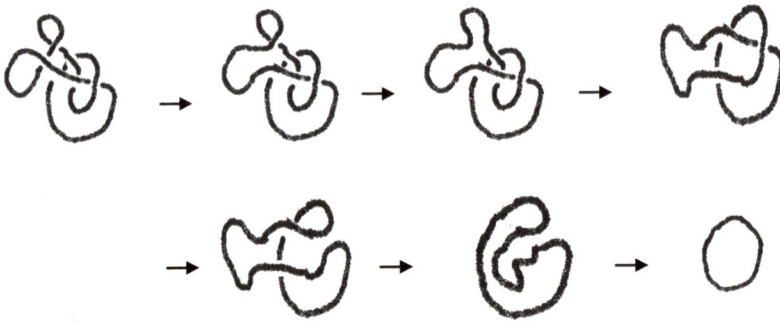

Fig.5.1.7

At first, we showed pupils a flowchart on deformations of a knot and how to draw these diagrams. Then we showed three types of *Reidemeister moves* (Fig.5.1.2) and confirmed that every deformation is one of the *Reidemeister moves*.

(2) Let's draw a flowchart on deformations of a knot.

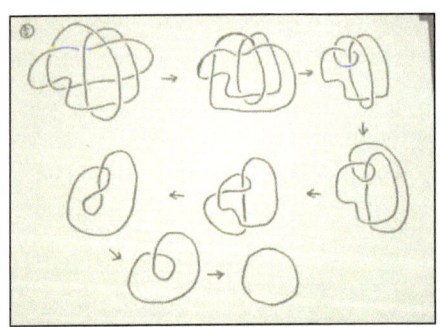

Photo 5.1.3(a): A pupil's work

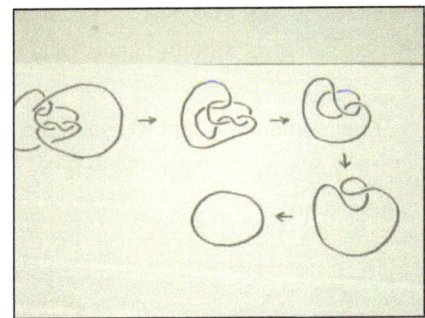

Photo 5.1.3(b): A pupil's work

The 3rd period

(1) Let's find a non-trivial knot.

⇒ Look at this knot. This is a *DNA knot*.

Can you untie this DNA knot?

How do you find whether this knot untie or not?

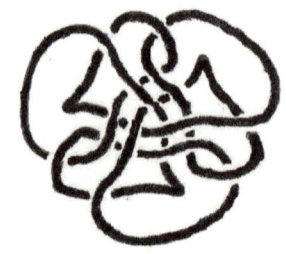

Fig.5.1.8

It is difficult to find whether this knot is untied or not. Then we get the pupils think about simpler knots.

⇒ Let's compare three couples of knots.   Which ones you can untie?

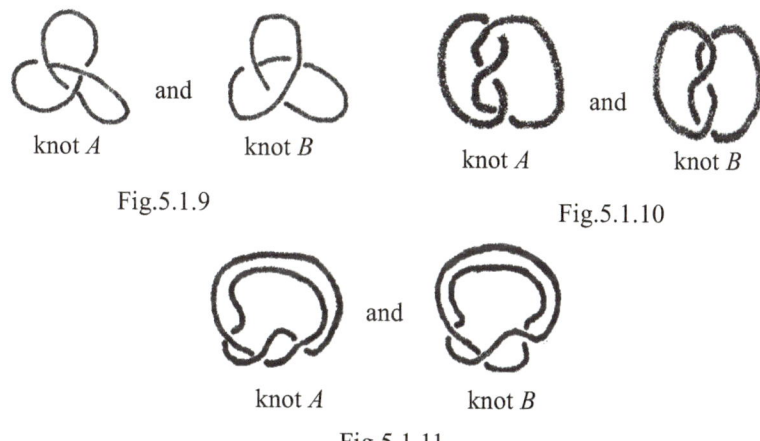

knot *A*   and   knot *B*

Fig.5.1.9

knot *A*   and   knot *B*

Fig.5.1.10

knot *A*   and   knot *B*

Fig.5.1.11

The knot *A* in each pair of knots in Figs.5.1.9-5.1.11 is a non-trivial knot.

⇒ Let's color each arc of the knots using one of the three different colors: red, blue and yellow.

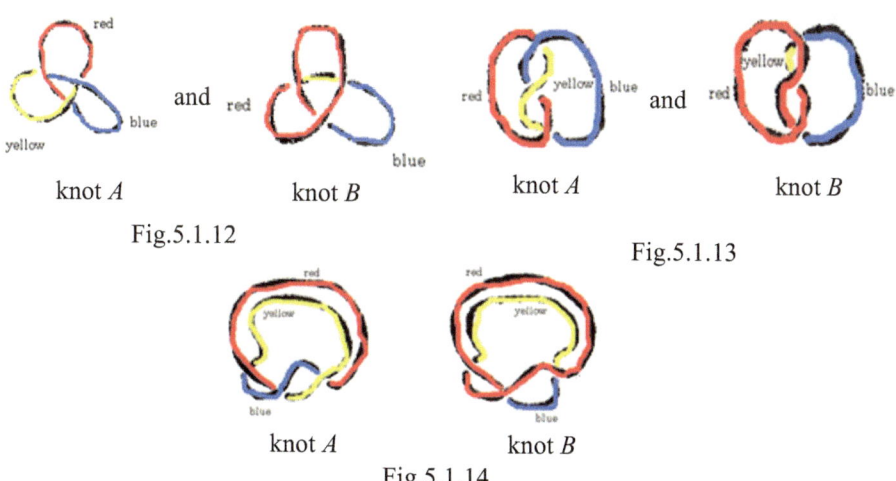

knot *A*   and   knot *B*

Fig.5.1.12

knot *A*   and   knot *B*

Fig.5.1.13

knot *A*   knot *B*

Fig.5.1.14

⇒ Let's find a difference between the knots in each pair.

In the knot *A* in each pair, it is possible that three different colors come together at each crossing. This is a characteristic of a non-trivial knot. Thus, we see that if we can color on each upper arc of a knot so that three different colors come together at each crossing, the knot may be a non-trivial knot.

The 4th period

(1) Let's make a *tri-colorable* knot.

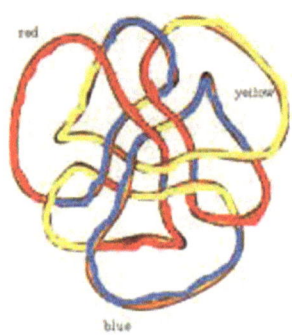

Pupils made some knots freely and colored each arc of the knots using different three colors.

Fig.5.1.15

Impressions of pupils   The results of a survey on these lessons (for the 152 pupils in the 3rd grade of junior high school) are the following:

(1) Check following items

   Very interesting.......................26%

   Interesting...........................59%

   Not so interesting...................13%

   Don't like this....................... 0%

   No reaction 2%

(2) What part were you interested in ?

   · Coloring each arc of some knot using different three colors

   · Finding untied knots by coloring

   · Analogy to a puzzle

   · Making a knot using a string

   · Writing a knot on a paper

   · Untying a knot using a string

   · Mystery of tri-colorability

   · Drawing diagrams of knots

   · Reason why three colors are used

   · Analogy to a play

   · Mathematics of knot theory

Photo 5.1.4

(3) What part did you feel difficult to understand this knot lesson?

   · Why are three colors used ?

   · Is this lesson mathematics?

   · Only looking is needed in this lesson?

   · In what case we can untie a knot?

   · Meaning of coloring each arc of a knot by using different three colors

· Why we cannot untie any tri-colorable knot?
· Relation between the Reidemeister moves and a deformation of a knot
· What is a conclusion in this lesson?
· Are you deceive me?

(4) What part do you feel that this is mathematics?

· Laws
· Logic
· 3-dimension
· Atmosphere in knot theory
· Result of each problem
· Open problem

Photo 5.1.5

5.1.4 Teaching plan (2)   The main theme of the second teaching plan is the *linking number*. In the practice here, we experimented with the 2nd grade pupils in the junior high school. The lesson plan consisted of the following 4 periods:

The 1st period: Which *link* is untied? (1)

The 2nd period: Which *link* is untied? (2)

The 3rd period: Investigate other links

The 4th period: Linking number

Although we could have a lesson to the second period, we have not yet experienced lessons to the 3rd and the 4th periods. This report is based on the 1st and 2nd periods.

The 1st period

(1) Look at two types of links.

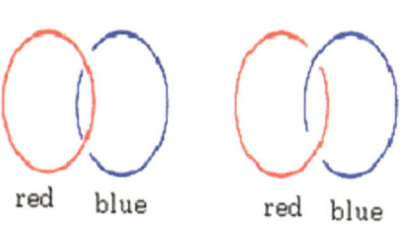

Fig.5.1.16

Photo 5.1.6

⇒ Let's deform these links. Then can you find which link is untied?

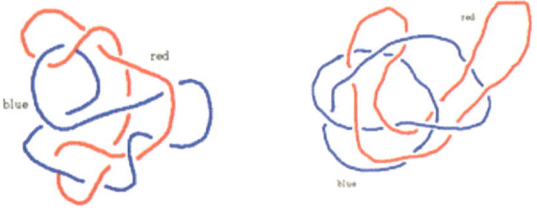

Fig.5.1.17

(2) Let's find a way to see an untied link on a paper. It seems difficult for us to find an untied link. Let's think of simpler links described as follows:

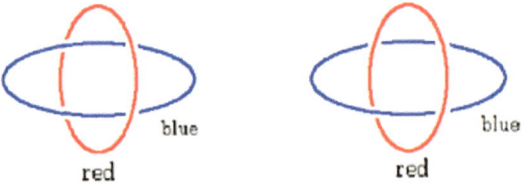

Fig.5.1.18

The left side link is an unlink. It has 2 crossings with a red arc as an upper arc, and the right side link has 3crossings with a red arc as an upper arc.

⇒ Let's think of another two links described as follows:

How can you find any diagrams of these links on a paper?

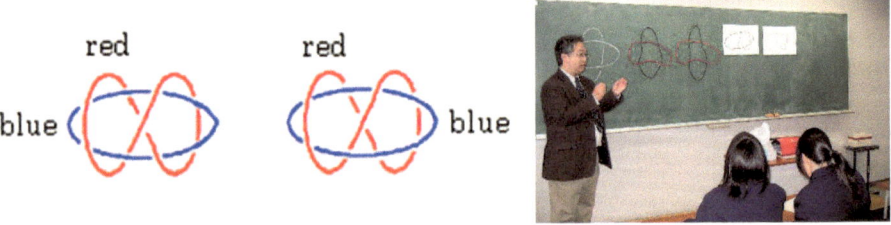

Fig.5.1.19   Photo 5.1.7

(3) What is your *conjecture*?

The pupils' *conjecture* is the following:

"If a link has even crossings which caused from that one arc is over another arc, then it is untied, and if a link has odd crossings which caused from that one component is over another component, then it is tied."

(4) Let's deform the two links above by using strings and examine your conjecture.

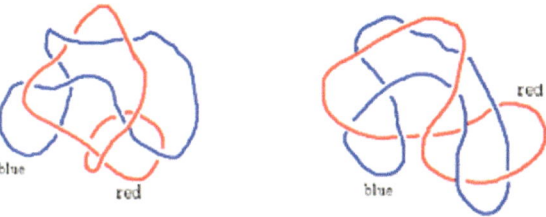

Fig.5.1.20

The left side link is the same as the link of Fig.5.1.21, and the right side link is the same as the link of Fig.5.1.22.

Fig.5.1.21   Fig.5.1.22

Most of the pupils got the same conjecture by using some other links. They thought that their conjecture is perhaps true.

The 2nd period

(1) Confirm the pupils' conjecture by using two links:

"If a link has even crossings which caused from that one arc is over another arc, then it is untied, and if a link has odd crossings which caused from that one component is over another component, then it is tied."

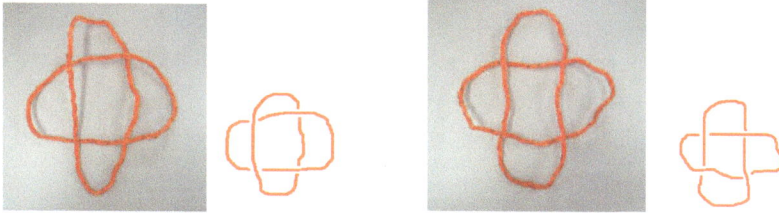

Photo 5.1.8: "An untied link" (left) and "a tied link" (right)

The pupils found that their conjecture is not true generally.

(2) Find another way to see a difference of these links.

⇒ Let's think of *oriented links* described as follows.

Link *A*   Link *B*

Fig.5.1.23

⇒ Look at every crossing on the oriented link.   How many types of crossings can you find on these links?

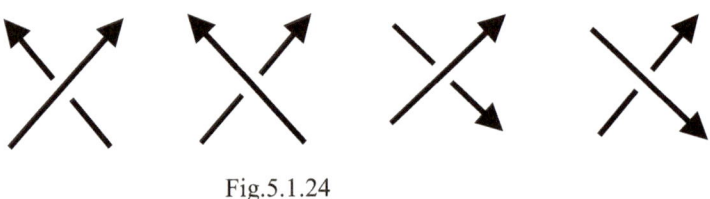

Fig.5.1.24

We see that there are only the two types on the crossings.

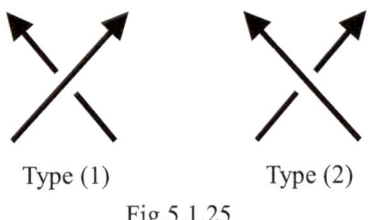

Type (1)   Type (2)

Fig.5.1.25

⇒ How many crossings of type (1) are there in the links? How many crossings of type (2) are there in the links?

In the Fig. 5.1.23, the right side link has 2 crossings of type (1) and the left side link has 4 crossings of type (1).

⇒ Let's find a way to see a difference in two links.

Some pupils found a way to see a difference in two links of the figure 5.1.23 as follows:

"In the right side link, the number of type (1) = the number of type (2), and in the left

side link, the number of type (1) ≠ the number of type (2). "

The 3rd period (This practice is not yet tried.)

(1) Confirm the pupils' *conjecture*

In the 2nd lesson, they made a conjecture as follows:

"In the link *A* of Fig. 5.1.23, the number of type (1) = the number of type (2), and in the link *B* of Fig. 5.1.23, the number of type (1) ≠ the number of type (2) ."

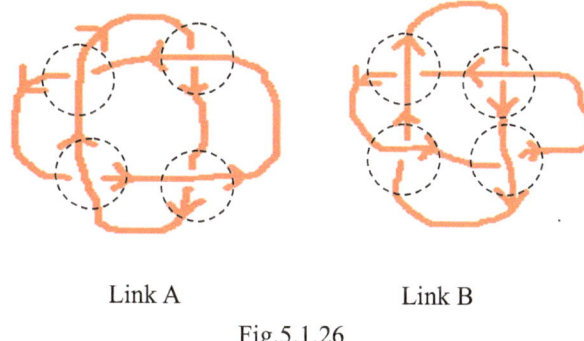

Link A   Link B

Fig.5.1.26

⇒ How about the inversed oriented links? Let's examine that this conjecture is true for the opposite oriented links.

This conjecture is also true for the opposite oriented links.

(2) Let's make some other links by using strings and examine this conjecture for the links.

The 4th period (This practice is not yet tried.)

(1) Let's count +1 for a crossing of type (1), and -1 for a crossing of type (2) and take the sum of ±1 over all the crossings in the link.

⇒ Let's deform a link into any other link and examine the sum of ±1 over all the crossings in the link.

⇒ What do you find about the sum?

The same link takes the same absolute value.

(2) What is your conjecture?

Thus, the pupils can understand "linking number".

Impression of pupils   After the 2nd lesson, we had a *questionnaire* to the pupils. The question is as follows:

(1) Check the following items   low   middle   high

  · Inclination   1 - 2 - 3 - 4 - 5

  · Understanding   1 - 2 - 3 - 4 - 5

  · Interesting   1 - 2 - 3 - 4 - 5

  · Wonder   1 - 2 - 3 - 4 - 5

  · Do you have many questions   1 - 2 - 3 - 4 - 5

  · I want to study more about the knot   1 - 2 - 3 - 4 - 5

  · I understand meanings to study mathematical knots   1 - 2 - 3 - 4 - 5

(2) Conclusion

The results of the questionnaire on these lessons (for the 116 pupils in the 2nd grade of junior high school) are as follows:

· Inclination

1 (5),   2 (7) ,   3 (29) ,   4 (40) ,   5 (19)

· Understanding

1 (8),   2 (14),   3 (32) ,   4 (27) ,   5 (19)

· Interesting

1 (8) ,   2 (11) ,   3 (34) ,   4 (31) ,   5 (16)

· Wonder

1 (7) ,   2(8) ,   3(28),   4(35),   5(22)

Photo 5.1.9

· Do you have many questions

1(7),   2(11),   3(30),   4(30),   5(22)

· I want to study more about the knot

1(8),   2(19),   3(44),   4(19),   5(10)

· I understand meanings to study mathematical knots

1(10),   2(18),   3(47),   4(15),   5(10)

(3)   Pupil's comments

< Positive comments >

  · I think that it is useful to study DNA.

- The knot is very popular for me and I'm very interested in the knot lesson. I feel wonder about the knot.
- I want to know the reason why the linking number is an invariant?
- The knot lesson was very difficult, but I am very interested in the knot lesson.
- I am interested in finding some laws.
- I feel that we need a deep thinking to understand the knot.
- I was interested in the 2nd period lesson.
- I need more times to study, but I am interested in this lesson and I can understand some contents of the knot lesson.
- Though I did not pay much attention to your lecture of knot lessons, a knot is very interesting to me and I thought of a knot by using a string and by trial and error.
- Since I took the knot lesson for the first time, I had many questions in this lesson. However, this lesson is very interesting.
- Though it is very difficult for me to understand the 2nd period lesson, I am interested in the knot world.
- Though the last part of the 2nd lesson is difficult for me, I want to study about the knot more times.
- Though I never thought so far whether a knot unties or not, I am interested in the existence of a law to distinguish some knots.
- I understood gradually the meaning of the contents in the knot lesson.
- Because I could wonder with only a string so much, I feel that the mathematics is a more mysterious thing.
- I think that the person who found out a law on the oriented links is great. The knot lesson is the mathematics that I can feel the reason for being. I was interested in practicing myself by trial and error.
- Though I understood easily a relation between an even number and an odd number, I did not understand a relation of arrows.
- It was different from usual mathematics lessons and we did not use numbers very much. I could understand easily the contents and I could feel about them so good by using my brain very much.
- If I saw only a figure of a knot, I had never understood the meaning. However, since I thought of the knot by using a string by trial and error, I understood the feeling.
- It is easy to understand the knot lesson.
- Though a knot is popular to us, I have never known that there are many mathematical wonders in the knot. Since, I think, there are many wonders in our world, I would like to have interests in everything of our world.

- I could not understand a knot if I see it only once. It is interesting to understand it by finding a law.
- It is interesting that there are many ways to untie a knot.
- At first, I could not understand the meaning of ⤢ and ⤡ , but after I knew the meaning, it was interesting to me like a puzzle. I could understand many things on the knot.
- We want to know only whether a knot unties or not, but we need a deep thinking to know it.
- At first, I could not understand the way to calculate the sign of every crossing by using the arrow. But when my friend taught me the way, I could understand the way and I felt good.
- I feel that the knot is very popular for us but we need a deep thinking to undertand the knot. I feel that this is a very difficult theorem and think that it is difficult to prove this theorem.
- I feel good finally.
- I have never taken these classes on before and we feel freshness to the knots. They are interesting.
- It was mysterious at all to understand whether a string unties or not by the crossing rules. I thought that the person who discovered such a thing was great.
- I want to take other classes like this knot lesson.
- The knot lesson was complicated, but when I understand the knot, I feel good.
- I had a feeling that is different from the normal mathematical feeling.
- It was hard to me to follow the knot lesson but I enjoyed the knot lesson by playing on the knot. I like the ring of wisdom in Japan, and the link is like it.
- I was interested in the lesson that a teacher teaches mathematics using the knot. I enjoyed the class and I need those lessons.
- It was the first time for me to take a mathematics class without calculations. Fresh for me!
- The 1st period is very interesting and the 2nd period is difficult for me.
- I felt that I use my brain after a long absence. I concentrated too much on it and was tired.
- I was interested in finding a rule.
- I felt that we are learning on very deep mathematics. I had a sense of incongruity about that popular knots are concerned with mathematics, and felt good adversely.
- There were many difficult problems in the knot class, No matter how even if it was

complicated, it was interesting that the opening was the same.

- I thought that the flexible idea that found a law by looking from various viewpoints was important.
- I would understand it later though first I could not understand what I did. It was interesting
- I used an actual string. I felt the knot strange and was interested in it. In addition, I thought that I wanted to have these classes sometimes if there was an opportunity.
- There were many wonders on the knot lesson and I was interested in the knot a little. On the other hand, I had a question "Is this mathematics?"
- It was good that I had interests to know about geometry more. Only two times practices were done but I was very interesting.
- I did not know why these lessons are mathematics. I was weak in mathematics, but I could understand them.
- I did not understand the explanation on changes of an arrow well and was puzzled. Though I thought that at the beginning these lessons were easy, I thought that it was complicated and difficult whereas we had thought of only two strings.
- It was pleasant that I simplified the figure of a complicated string in a head. It was easy to understand that I drew it, but thought that my fun decreased.
- I thought that if I knew the knot more, then I could understand what kind of knot I can untie.

< Negative comments >

- It was complicated to find the difference between ⤡✕ and ⤢✕ .
- No comments.
- I felt obediently that I liked a problem to solve a formula than such a class.
- I have gradually lost it toward the last. I am disappointed.
- To be frank, I did not understand it well what wanted to do it. I was not interested in the knot lesson.
- I did not understand and it was difficult to me.
- Because it was very difficult to understand, I could not understand it very much. Because I am not proud of thinking about these contents, I was confused very much.
- It was delicate.
- We found a way to untie a knot, but the other knot was not untied by this way.
- No answer.

5.1.5. Conclusion

On teaching "tri-colarability"   These knot lessons were different from usual lessons and the pupils felt something new. They had the same starting point. Then we saw that many pupils were interested in these classes. The knot is popular for the pupils since every pupil have ever untied a string. Since we cannot use a formula, there were many pupils who tried to find some rules in a puzzle sense. We could not understand whether a knot unties or not only by having looked at the first sight. Many pupils were interested in the existence of a method to examine what I could not understand only by having looked at the first sight. Since we omitted the proof of the tri-colorability, the pupils had some questions after the knot lessons. We think that many pupils thought that our purpose was untying some knot. What we can understand by the tri-colorability is that we cannot untie the knot, and is not that we can untie the knot. If the purpose of the pupils is untying a knot, we think that they could not understand the meaning and did not feel usefulness of the tri-colorability. Instead of finding whether a knot unties or not, we think that we need a class finding the *same knots*. We think that many pupils did not understand the way to distinguish knots by coloring on a knot. It seems to be hard to mention these contents in the knot lesson as mathematics other than saying it based on a textbook. However, about 70% pupils felt these contents in the knot lesson mathematical. We felt the possibility as new teaching materials in the knot lesson. The pupils did not know any theorem and they made "their theorems" in theses knot lessons. There were many students who felt difficulty, because they must think from the beginning. There is not always a well known solution, and on study there is a case they could not even find one answer. About various problems on our life, we do not know whether the number of the answers is one or not. We think that there were the pupils who felt that mathematics included such a part through these lessons. In this class, we made many "interesting mathematics" for the junior high school students, so that the conclusion of the lessons became vague. Thus, we think that there are many pupils who felt these lessons difficult.

On teaching "linking number"   In the 1st period pupils found their conjectures as follows:

"If a link has even crossings which caused from that one arc is over another arc, then it is untied, and if a link has odd crossings which caused from that one component is over another component, then it is tied."

We let the pupils think about which link is not untied gradually from simple models to more complicated ones. Through these lessons we could bring out this conjecture from the pupils. We used two types of links in the 1st period. For the links the conjecture stated above is true. Since this conjecture is a "if and only if" condition for the links, it was easy for pupils to find this conjecture. Many pupils were interested in the 1st period lesson. I think that it was good to have let pupils practice by using actual strings. In the 2nd period we needed oriented links. At first we tried that we let the pupils find putting an arrow on a string. However, it was very difficult and we think that we have not any meaning even if we guide it. Then we showed the pupils that we put an arrow on each string. We thought that we let them find a difference between two links. A few pupils found a difference between the two links. We think that if we had more time, more pupils should have been able to find such a difference. In the 2nd period, we used many prints for the pupils to practice on the knot lesson. We think that it was not good, because they have only a few thinking time by "*try and error*". We should give the pupils more thinking time by "try and error." We think that it is important for pupils to think about links by using strings. In this class, we do not think that the pupils found the oriented links and found an importance of differences between two links. These are very difficult for us and we seemed to be accidental. We think that the pupils should consider on links using oriented links. Finding the difference between $\times$ and $\times$ are very important for us to understand the 3-dimensional space. We need to show a solution of their conjecture. For this, we need "Reidemeister moves". However, we do not try these practices in the junior high school. These lessons were over a halfway and we think that the pupils had some vague feelings. We should have more times for these practices.

References

[1] A. Kawauchi et al, Knot Theory (in Japanese), Springer Verlag, Tokyo, 1990. (English expanded version: A Survey of Knot Theory, Birkhäuser Verlag, 1996)

[2] C. C. Adams, The Knot Book, W.H. Freeman and Company, 1994. (Japanese version translated by T. Kanenobu, 1998).

[3] S. C. Carlson, Topology of Surfaces, Knots and Manifolds, John Wiley & Sons, Inc., 2001 (Japanese version translated by T. Kanenobu, 2003)

[4] A. Kawauchi and T. Yanagimoto et al, An Approach to Teaching Knot Theory in School Mathematics for Pupils and Students (in Japanese), Project of Teaching Knot Theory in School Mathematics, Research Report as Educational Action in 21st Century COE Program "Constitution of wide-angle mathematical basis focused on Knots(Osaka City University)" Vol.1 2005, Vol.2 2007, Vol.3 2009.

[5] K. Iwase, A Study on Teaching Mathematical Knots, Bulletin of the Tennoji Junior & Senior High School Attached to Osaka Kyoiku University, No.47, 2005.

[6] T.Yanagimoto, K. Iwase, M. Terada, H. Kanaya, N. Masuda, T. Shimonosono, "A study on Teaching Knot Theory in Schools", TSG4:"Secondary Teaching and Learning" of ICME - EARCOME3, 2005.

[7] K. Iwase, Mathematical Links as Teaching Material for Junior High School Students, Bulletin of the Tennoji Junior & Senior High School Attached to Osaka Kyoiku University, No.51, 2009.

[8] T. Yanagimoto, Y. Seo, K. Iwase, M. Terada and R. Kaneda, An Approach To Teaching Knot Theory in Schools, Proceeding of the 4th East Asia Regional Conference on Mathematics Education (EARCOM4), 411-416, 2007.

[9] T. Yanagimoto, Y. Seo, K. Iwase, "A Study on Making Knot Theory into a Curriculum", Proceeding of the 5th East Asia Regional Conference on Mathematics Education (EARCOM5), 637-644, 2010.

5.2 Education practice in public junior high school

In this section we look at how we came to have the knot theory lessons in the public junior high schools. Public junior high schools in Japan have pupils who are good at mathematics and those who are not. There are also pupils who like school life and those who do not. We thought that knot theory lessons might be helpful in getting pupils interested in mathematics. Japanese public school teachers have to teach their classes in accordance with the course of study that has been decided under the national law. In order to have the knot lessons in public schools, we first had to obtain a permission from the Osaka Prefecture Board of Education and the principals at the schools. The public school math teachers did not know knot theory itself. In order to introduce knot theory to them, we have had monthly meetings since March 2009. 5 professors, 3 high school teachers and 8 public junior high school teachers participated in the meetings. Meetings were held for 3 hours on Saturday afternoon (Photo 5.2.1). The university professors were eager to have the lessons in the public junior high schools. In the meetings we discussed the lesson content, teaching methodologies and possible educational teaching aids.

Photo 5.2.1: Monthly Saturday afternoon meeting

With regard to the lesson content, we talked how to teach knot theory intuitively without mentioning its mechanism. We agreed that teachers should prepare their lesson contents which would interest the pupils and encourage them to think. We also agreed that teaching methodologies should assist the pupils in grasping the concepts intuitively. We discussed the methodology in which the pupils were directed to "twist", "slide" and "flip" strings in some lengths to intuit the theory. These three actions are related to *Reidemeister moves*. Various suggestions for educational aids such as a string, a soft wire and a tracing paper to step through the concept were made during the meetings. Worksheets would serve to help students transfer a knot in the 3-dimensional space into the plane. Instructive lectures on knot theory to the public junior high school teachers were mainly given by a high school teacher teaching at Tennoji Junior and Senior high school attached to Osaka Kyoiku University. He had already taught knot theory lessons to his junior high school pupils at his school. After 3 meetings, the 3 public junior high school teachers gave lessons on knot theory. Contents of the lessons are described in the following sections.

5.2.1 Education practice -1-   This lesson was carried out by R. Futahashi (Noda Junior High School, Sakai, Osaka) with 7 boys (6 in 2nd grade and 1 in 1st grade) as a club activity for 3 hours in May 2009.

Lesson Overview   The purpose of the lessons was to lead the pupils to find some rules to *untie a knot*. The teacher asked the pupils to try to find such a method and to figure out the best method to do so. Some pupils photographed all the steps of moves where a pupil considered in the process to untie the knot. The other pupils repeated the process and took photos of all the consecutive series of figures. After taking a number of series of photos, the pupils displayed them as a motion picture.

Lesson Details

(1) The teacher showed pupils a jumble of a tangled rope to get them interested in how to untangle it (Photo 5.2.2).

Photo 5.2.2: A jumble of tangled rope

(2) The teacher had shown the images of Photo 5.2.3, Photo 5.2.4, and asked the pupils to illustrate them on a paper. She had then explained that the *string* is joined like the one in Photo 5.2.5.

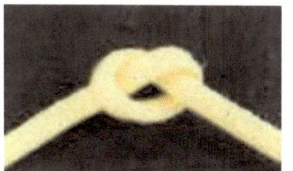

Photo 5.2.3   Photo 5.2.4   Photo 5.2.5

(3) The teacher then showed the pupils the figure in Photo 5.2.6 and asked them to think of the best way to manipulate the knot to make it simpler.

(4) The pupils were asked to explain how they changed the knot in their own words (Photo 5.2.6).

(5) The pupils displayed successive photos of knots (Photo 5.2.7).

"Move one part of the knot to move it appear as a simpler form."

Photo 5.2.6

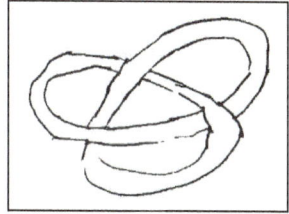

Photo 5.2.7

Pupil Reaction

(1) When the teacher showed pupils a jumble of a tangled rope to get them interested in how to untangle it and asked for opinions, one pupil suggested that it might untie a little if it were shaken around. Another pupil countered with a suggestion to try small, partial movements.

(2) The pupils focused on a *crossing point* to expose the upper crossing points before drawing the knot in papers (Fig.5.2.1).

Fig.5.2.1

(3)The pupils also tried to explain the process in their own words (Fig.5.2.2).

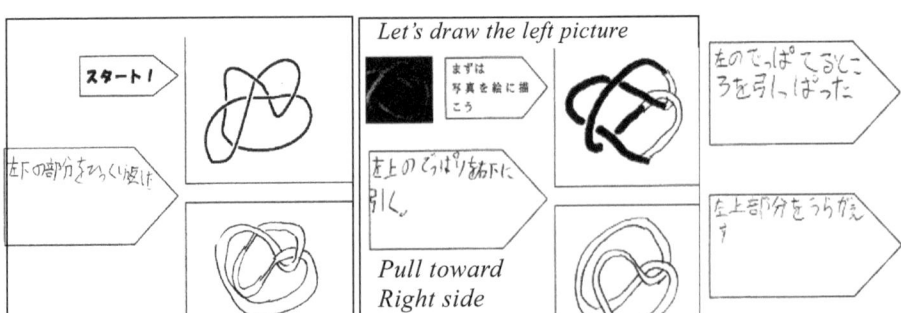

Fig.5.2.2

At this stage, the pupils did not express too much enthusiasm towards making drawings. They were more interested in actually untying the knot. They also expressed curiosities as to how far they could untie the knot and whether or not there were rules on how to untie them. None of them chanced on the idea of counting the number of crossing points.

(4) Instead, the pupils themselves turned the exercise into the game.

[The first game]

The first game in which they were interested required to untie the knot with the least number of moves. The pupil untying the knot with the least moves was considered the winner.

[The second game]

The second game they thought up was to take turns in untying a single knot. If a pupil could not proceed further when it was his turn, he was considered out and the next pupil would continue the process. The last pupil left was the winner.

The pupils continued to take consecutive photos as they played the games. They also started to spontaneously use words like "twist", "untwist", "pull" and "slide" to describe the string movements. They intuited all of the Reidemeister moves except for "flip".

(5) Eventually, the pupils found they did not need a physical knot to imagine the untying process and could draw the sequences on a paper directly (Photo 5.2.9). Instead of complaining that drawing the process is troublesome when they were in

the stage(3), pupils were eager to make sketches.

Photo 5.2.8: Describing on a paper

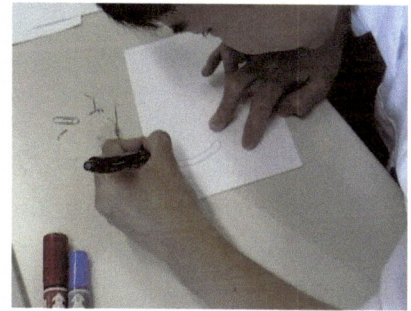

Photo 5.2.9

They could describe the knot quickly after having completed many activities. They even figured out a way to emphasize a moving part by coloring it in red(Fig.5.2.3).

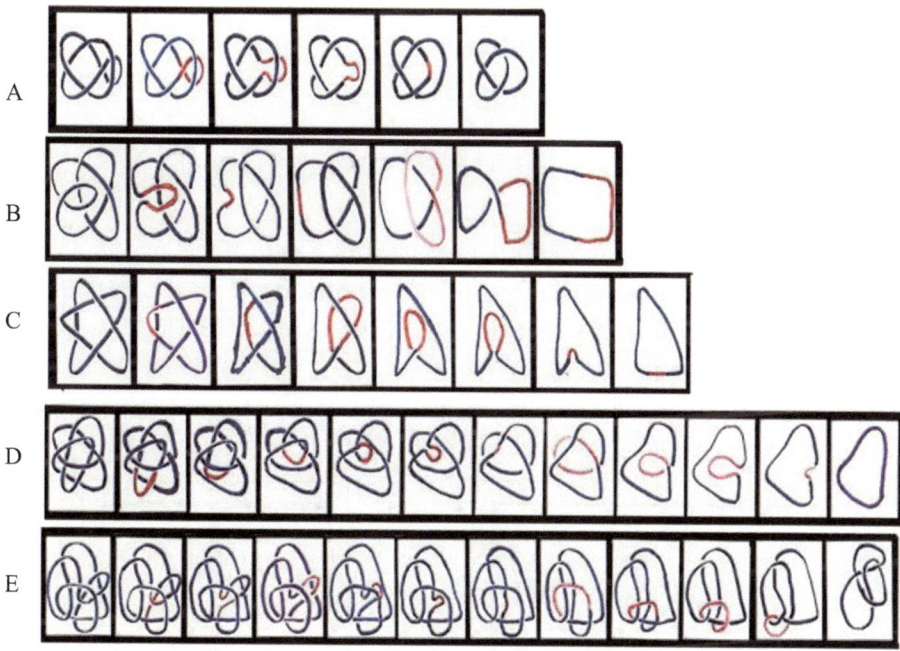

Fig.5.2.3

Questions posed by pupils.

- Can the knot be loosened?

- What is the least number of moves which the knot can be untied in?

- Is there a similarity between knots that cannot be loosened further?
- When pupils describe the knot in the plane, they ignored the length of the string to write it as simple as they could.

Pupils said that they disregarded the length of the strings when thinking of them. Furthermore, some said that it was as if they could imagine an invisible character untying the knot in a cartoon when they viewed the photos in motion picture (Photo 5.2.10).

Photo 5.2.10

Lesson Summary
- It was useful for pupils to image a movement of a part of the knot to untie it. Eventually they could image a process of untying a knot without an actual knot.
- The pupils could draw the knots in 2–dimension indicating which part of the string was over or under the other part.
- During the lesson pupils came to intuit that knots A and B and knots C and D were the *same knots*(Fig.5.2.4).

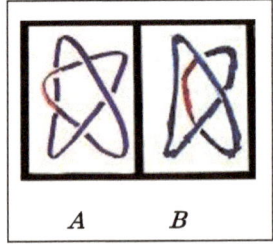
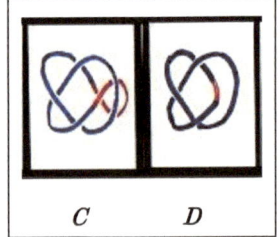

Fig.5.2.4

- Pupils noticed by themselves that they could simplify to untie the knot and there were models on untying the knots, because they had been given an enough time of activity.

- The pupils could grasp and appreciate a value of the 3-dimensional space through the lessons.

5.2.2 Education practice -2-   Two different classes were carried out by F. Arai (Tomioka Junior High School, Sakai, Osaka) as mathematics classes in July 2009. The first class was carried out with 36 in 2nd grade (18 boys, 18 girls) in two straight school hours. The second class was carried out with 37 in 8th grade (19 boys, 18 girls) in two straight school hours at the same day as the first class.

Lesson Overview   In the lesson of the first class, the teacher asked pupils how to *untie a given knot*. In the lesson of the second class, the teacher showed a knot and asked the pupils to draw the diagram as they see it from the opposite side of the knot.

Lesson Details -First Class-

(1) First, the teacher showed the pupils actual ropes as illustrated in Fig.5.2.5, Fig.5.2.6 and Fig.5.2.7. Next, she asked the pupils which ropes would become untied if both ends were pulled.

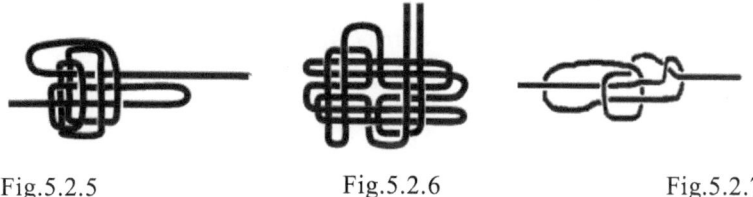

Fig.5.2.5   Fig.5.2.6   Fig.5.2.7

(2) The teacher pointed out the difference between the pictures of the knots A and B (Fig.5.2.8, Fig.5.2.9). Then she asked the pupils to draw their diagrams.

Fig.5.2.8: knot A Fig.5.2.9: knot B

(3) The teacher showed them the diagrams of the knots C and D (Fig.5.2.10, 5.2.11). Then she asked the pupils which part of the knot C should be moved in order to make the knot D. The teacher also asked her pupils which kind of a rule was used when the knot C is changed into the knot D.

82

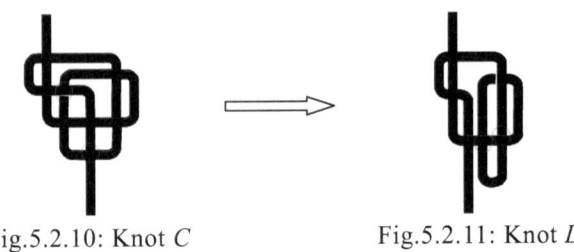

Fig.5.2.10: Knot *C*   Fig.5.2.11: Knot *D*

(4) The pupils tried to *untie the knots* in Figs.5.2.12, 5.2.13, 5.2.14   in order to find out what kinds of rules were used.

Fig.5.2.12   Fig.5.2.13   Fig.5.2.14

Pupil's Reaction of the first class
(1) The pupils were able to explain the process with words such as "twist" and "slide" which they used in their words (Fig.5.2.15).

Fig.5.2.15

(2) Although at first a pupil was unable to understand the content of the lesson presented in Fig5.2.16, eventually the pupil came to be able to solve the problems by alone as in Fig5.2.17.

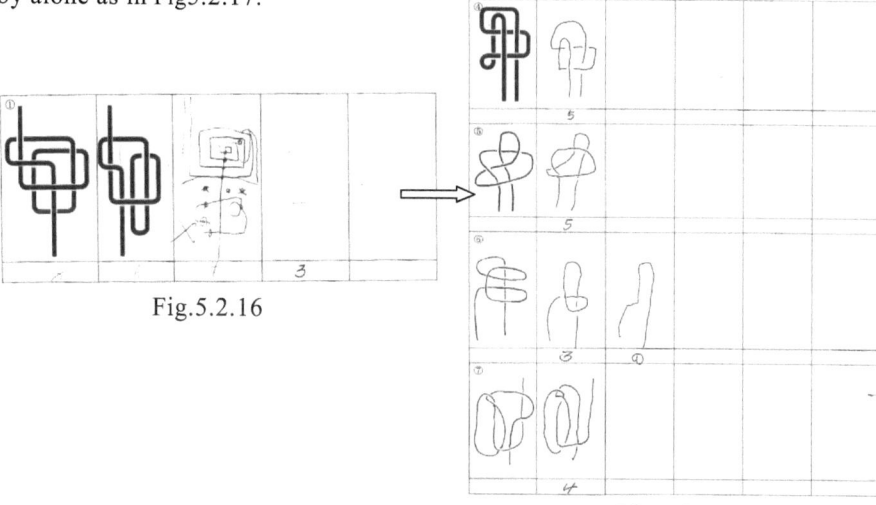

Fig.5.2.16

Fig.5.2.17

(3) A pupil who got better and better at untying the knots said with a confidence "Now I'm a real pro at this."

(4) The pupils checked each other on the order on untying the knot. Eventually, they found good procedures to develop a method to untie the knot.

(5) The pupils could think of the best way to manipulate the knot to untie it. The pupils could compare their processes on which part of the knot they untied. The pupils enjoyed untying a knot with an increasing speed.

Lesson Details -Second Class-

(1) The teacher asked the pupils to draw Japanese Kanji as viewed from the opposite side of the paper (Fig5.2.18).

(2) The teacher showed a string as it is shown in Fig.5.2.19. Then she asked the pupils to draw a picture of the string as seen from the opposite side of a string.

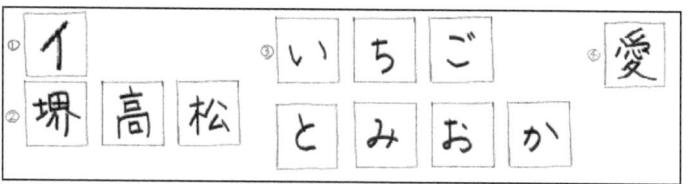

Fig.5.2.18   Fig.5.2.19

Pupils' Reaction of the second class

(1) The pupils could easily draw the picture of the characters as seen from the opposite side (Fig.5.2.18).

(2) When the pupils tried to imagine a figure of Fig.5.2.20 from the opposite side, most pupils thought that the correct figure was like Fig.5.2.21. Some pupils said that it was the same as the given figure. Then many other pupils tried to become eager to draw the correct figure.

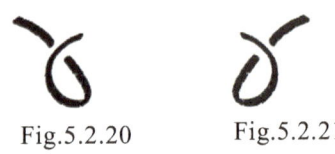

Fig.5.2.20   Fig.5.2.21

(3) The teacher gave a string to every pupil. After they made the knot using the given string (Fig.5.2.20) and saw it from the opposite side, some of them could draw the correct figure. Other pupils drew figures based on other imaginations first, and then checked whether or not their answers were correct by comparing it with the given figure.

(4) Some pupils could not accept the correct answer, and they began to discuss which figure was true.

(5) Some pupils could find a dimensional difference between the pictures of a Japanese character and a string. A pupil said "I actually feel a difference between the pictures of a Japanese character (Fig.5.2.18) and the string (Fig.5.2.19) seen from the opposite side".

(6) The pupils were surprised that the figure which they saw from the opposite side was the same as the figure which they saw from the front side (Fig.5.2.20).

Lesson Summary

-Some pupils asked the teacher to borrow a string to image an actual knot in 3 dimensions although the knot which was drawn on a paper could be seen easily. This fact told us that many exercises were needed to untie knots both in diagrams in the plane and in 3-dimensional figures.

-At the beginning stage of the lesson, an open knot was easier than a closed knot for the pupils to handle a knot, but a closed knot must be handled at some stage before finishing the lesson. Fortunately, the pupils were able to notice that any shape cannot be made unless a knot is closed.

-It was comparatively easy to express 3 dimensions with a string, and a string is a useful teaching aid easy to be used for pupils.

5.2.3 Education practice -3- This class was carried out by S. Ibaraki (Fifth Junior High School of Kaizuka, Osaka) with 36 pupils in 7th grade (16 boys, 20 girls) as mathematics classes in July 2009. The first class was carried out with 36 in 8th grade (18 boys, 18 girls) in two straight school hours.

A preliminary investigation   Several pupils cooperated with a preliminary investigation done to better understand the pupil's abilities to imagine the deformations of some knots. Our results were as follows:
- Most students could imagine untying a knot with a few crossing points.
- The pupils could correctly untie a knot with more crossing points. They untied the knot by looking at a sequence of pictures which were drawn by the pupils to untie the given knot.   A pupil who had a difficulty to *untie the given knot* in 2 dimensions asked to let the pupil use a string to simulate it.
- It was difficult for some pupils to untie a knot with additional crossing points without using a string.

Lesson overview   The teacher made his lesson contents according to the first result of a preliminary investigation.
- It was found that each pupil could not untie the given knot easily, so that he decided to give his pupils a knot which had only a few crossing points first in his lessons. Next, he gradually gave a knot with more crossing points.
- In addition, at the beginning of his lesson, his pupils could use the educational aids such as a string and a soft wire. They used the aids to *untie the given knot* which was drawn on a paper.
- It was decided that his pupils used words like "twist" "untwist" "pull" also "slide" or "flip" to describe the string's movement. His pupils found that they could untie the given knot easily when they used words.

Lesson Details -First lesson-
(1) Think of the knot (Photo 5.2.11) whether it is untied or not. This is an objet d'art in front of the gate of a public athletic field, which was built in honor of the OSAKA EXPO in 1970.
(2) Express their opinions on how to deform the objet d'art into a circle.
(3) Confirm that it is untied by using a picture   taken from another angle (Photo 5.2.12).

86

Photo 5.2.11   Photo 5.2.12

(4) Think of the knot whether or not it is untied by using a string (Fig.5.2.21).
    Think of the knot whether or not it is untied by using a soft wire (Fig.5.2.22).

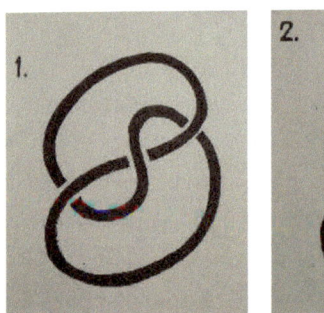 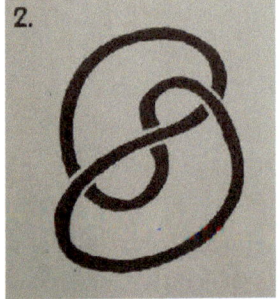

Fig.5.2.22

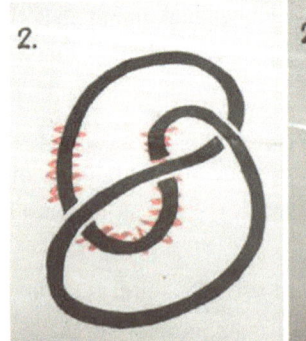 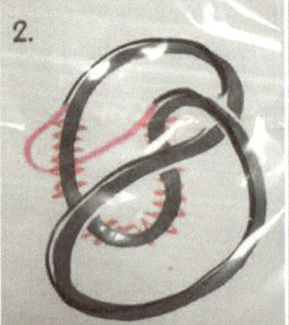 

Photo 5.2.13: Process of untying the knot using a tracing paper

(5) Draw and explain a process of untying the knot (Fig.5.2.23) before trying by using an actual knot.

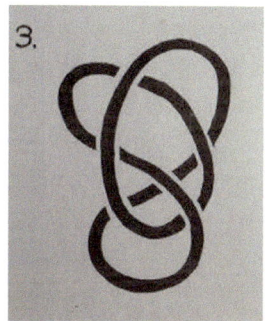

Fig.5.2.23

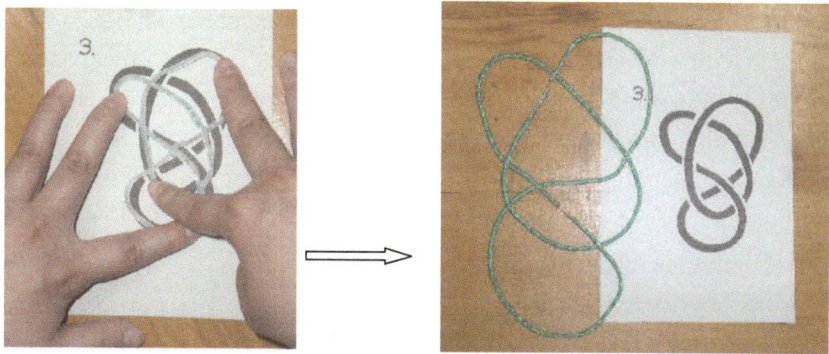

Photo 5.2.14

Lesson Details -Second lesson-

(1) Judge and explain a process of untying the knot before trying by using an actual knot. (Fig.5.2.24)

(2) The next trial was to write a footprint to the given knot. Here, the footprint means that the pupils could not move the arc containing the footprint to untie the knot. Imagine whether or not the given knot can be untied and draw a few consecutive figures until the knot is untied (Fig.5.2.25).

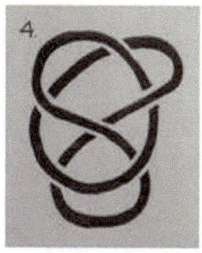

Fig.5.2.24

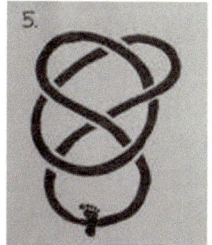

Fig.5.2.25

88

(3) Comparing the given knot, the consecutive figures and the untied figure, the pupils answered how to move the knot and how they thought of untying it.

Pupils' Reaction

On (1)-(3)of the first lesson: When the teacher showed a picture of the objet d'art to his pupils, the pupils recognized the athletic field's name easily and they were interested in the picture. After that they made a similar knot of the picture using a soft wire. The teacher asked his pupils how to move the given soft wire to untie the knot seen in the picture.

On (4) of the first lesson: During the pupils' trials of the knots of Figs.5.2.21 and 5.2.22, they could draw the knot on a paper by manipulating the actual string in 3 dimensions. On the contrary, they could recreate a similar knot using an actual string in 3 dimensions while looking at the figures of Figs.5.2.21 and 5.2.22.

On (5) of the first lesson: When the pupils used tracing papers to draw a sequence of figures, some pupils had a difficulty to draw the figures. However, they could untie the given knot using a soft wire. Many pupils draw 3 or 4 figures until they untied the given knot.

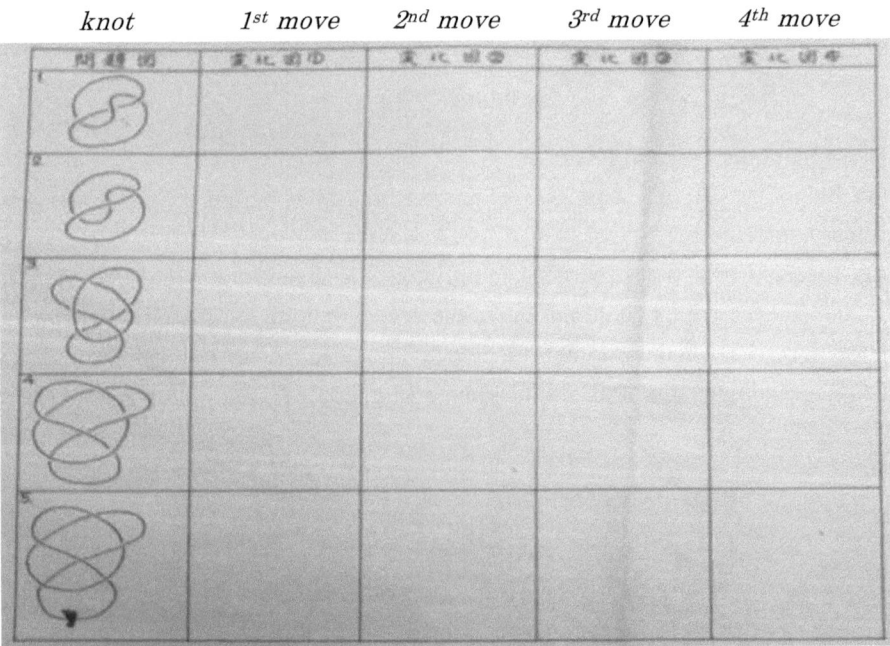

Fig.5.2.26: Work sheet used by every pupil

Lesson Summary

(1) At the beginning of the lesson, the teacher showed a picture of the objet d'art. It helped pupils become interested in this lesson, because untying a knot was not familiar to the pupils.

(2) The pupils mainly used two kinds of educational aids, a soft wire and a string.

- By using the soft wire, it was rather difficult to recreate a figure similar to the given knot which was drawn on a paper. It was easy to watch the consecutive changes when the pupils untied the given knot.

- The string was not suitable to watch the untying process. It was easy to make a figure similar to the given knot which was drawn on a paper. It was important for the pupils to check which arc was over or under the other arc when they manipulated a crossing point. They could correctly check and realize the crossing point.

(3) "Kitchen board" is useful to make the pupils the same experience each other. (Photo 5.2.15)

(4) The teacher advised his pupils to use the words like "twist", "untwist", "pull", "slide" and "flip" to describe the string's movement. They helped them manipulate the given knot. These words were not given by the teacher, but the pupils themselves began to use them in their own words while they manipulated the knot.

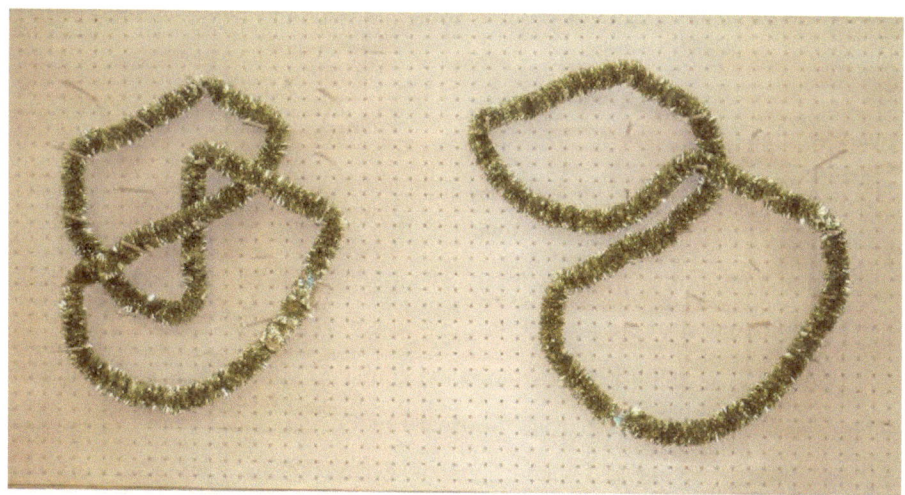

Photo 5.2.15: "A kitchen board"
Insert pegs for hanging around 3m closed soft wires

90

(5) The pupils experienced the following three stages during the lessons.

First level: How to manipulate physically a knot.

Second level: How to draw graphically a knot.

Third level: How to manipulate mentally a knot.

The teacher found that it was important for his pupils to go through and review the 3 stages many times for reinforcement.

(6) The teacher noticed that his pupils could easily flip the string towards them, but they had a difficulty flipping them the other way. To overcome this difficulty, it was useful for the pupils to go through and review the 3 stages of (5).

Teacher's After Class Discussion 9 math teachers had discussed this lesson after the class. 5 teachers were from 5 public junior high schools in Kaizuka and 4 were our study group members. Our conclusions were as follows:

- It is useful to use a tracing paper for pupils to draw and explain a process of untying the knot. While using it, they can raise a teaching level from manipulating an actual knot to draw a knot to manipulating mentally a knot. For some pupils, it is useful to go through and review the 3 levels many times for reinforcement.

- "A kitchen board" is useful to make the pupils the same experience each other (Photo 5.2.15).

- Pupils had a difficulty flipping a string the other way. To be overcome this difficulty, the teacher should prepare something for the pupils' thinking of the knots of Figs. 5.2.24 and 5.2.25.

5.2.4 Summary

(1) We have had monthly meetings for around three hours on Saturday afternoon since March 2009. We discussed to what degree and with which methods we could teach pupils on knots. We also discussed terminologies and materials we would use in conducting the knot lessons.

(2) The knot lessons have not included the course of study that has been decided under the national law. However, the Board of Educations of Osaka Pref., Sakai City and Kaizuka City and school persons where the three teachers work were willing to support them.

(3) In public junior high schools, there should be various pupils who are eager in mathematics or not earnest. In addition, most of teachers did not know knot theory

itself. Thus, we were first afraid that the knot lessons in public junior high schools would not succeed.

(4) The contents of the three teacher's lessons were going to let the pupils discover the basic transformations of a knot, so-called Reidemeister moves. The pupils intuitively used the moves in their own words while they manipulated the knot.

(5) The educational aids such as a string, a soft wire and a tracing paper are useful to step through the knots and to bring up pupil's mathematical thinking. Worksheets served to help pupils transfer the knots from the 3-dimensional space to the plane and manipulate mentally the knots.

(6) It is important for the pupils to check which arc is over or under the other arc when they manipulate a crossing point. This check is useful to let pupils study in knot lessons. This check is useful to grow up a space perception and a 3-dimentional sense.

(7) The reasons why the three teacher's lessons succeeded were because they asked suitable questions for supporting their students. In addition, we had monthly meetings for supporting the three teachers with 5 professors, 3 high school teachers and 5 public junior high school teachers.

An impression of classes on mathematical knots in public junior high schools
In 2010, I attended two public classes on mathematical knots in public junior high schools, a 8th grade class in the Sakai City Tomioka Junior High School, done by Arai on June 15 and a 7th grade class in the 5th Kaizuka City Junior High School, done by Ibaragi on July 3. In a public junior high school, there should be various students who are eager in mathematics or not earnest. I was interested in how a teacher tells a knot to such students since I received announcements of the classes. A knot is a phenomenon to hide behind in nature basically and is found in various sciences like DNA studies today. I believe that it is a very important education today to let pupils learn a knot in an early stage. Both the classes where I attended were classes soon after lunch. Some pupils seemed to be sleepy at the beginning of the classes. Because introductions of the teachers were good, all the pupils were absorbed in a knot gradually, and the classes became cheerful until around the end of the classes. I was glad that the pupils had found interests in knots. The contents of lessons were going to let the pupils discover basic transformations of a

knot, so-called Reidemeister moves. It is an advantage that 3-dimensional space is expressed by just showing a crossing of a knot. I have the impression that these lessons could grow a space perception and a three-dimensional sense even for the pupils not so good at mathematics. In addition, I felt that a knot could serve as a teaching material to grow a space perception and a three-dimensional sense in all school years of the public junior high school. We meet a knot approximately every day. I am convinced that if we know a meaning of a knot once in childhood, since then a knot will be helpful to send a creative way of life with a space perception and a three-dimensional sense. This educational challenge was probably practiced for the first time in the public junior high schools of Japan. I express a respect for the school persons supporting it and the teachers Futahashi, Arai and Ibaragi. (This note is Kawauchi's report at the meeting of High School and Osaka City University Mathematical Cooperative Association, held in November 7, 2009.)

An approach to teaching knot theory in public junior high school mathematics
I attended a fifth grade class at Tennoji Elementary School attached to Osaka Kyoiku University in 2004, where a knot lesson was held for the first time. I had never attended the knot lesson before. Professor and Mrs. Wittmann who came from Germany attended the lesson. I remembered that he had a highly appreciation of this lesson. I thought that the pupils were interested in the lesson and it was useful to bring up their mathematical thinking. Next, I visited a 3rd grade class at Tennoji Junior High School attached to Osaka Kyoiku University (=:TJHSAOKU), where a knot lesson was held, too. In the lesson, the pupils manipulated knots drawn on a paper to consider which knot is untied or not. They enjoyed the lesson.

The knot lessons have never been taught at the public junior high schools in Osaka Prefecture although pupils had been taught at TJHSAOK. Okamori asked Terada and me to have the knot lessons at public junior high schools in February 2009. So, we organized the Public Junior High School Knot Lesson Conference. The members were 5 university professors, 3 high school teachers, 8 public junior high school teachers. We have held the meetings of the Conference in Saturday afternoon every month since March 2009. Through the meetings, Iwase explained the public junior high school teachers about the mathematical education of the knot, the purpose and meanings of having knot lessons in school, and the knot lesson's records done at TJHSAOK, etc. After that, we carried out the knot lessons by the three teachers at three public junior high schools. The records are included in this chapter. I have already reported our projects as follows:

The knot lessons at TJHSAOKU in the proceedings was reported of the Journal of Osaka Public Junior High School Mathematical Education Conference, 10 (2006), 41-44. The contents of the three teacher's knot lessons were also reported by Arai, Terada, Kawauchi and me at the meeting of the High School -Osaka City University Mathematical Cooperative Association held on November 7, 2009.

I was sincerely impressed that the pupils had found interests in knots in school mathematics. In addition, I myself was very glad to be able to learn knot theory through the monthly meetings. I express my sincere thanks to our conference members, especially Emeritus Professor H.Okamori and Professor A.Kawauchi. (This note is Kamae's report.)

Note: This section was carried out through the discussion by the following members of the conference on knot lesson.

Fumiko Arai	Ryonan Junior High School, Sakai City
Akio Fujiwara	Katashimo Minami Junior High School, Kasiwara City
Rumiko Futahashi	Noda Junior High School, Sakai City
Toshihiro Homma	Emeritus Professor of Kobe Shinwa Womens University
Satoshi Ibaraki	Tennoji Junior and Senior High School attached to Osaka Kyoiku University
Haruo Inui	Uenomiya High School
Ken-ichi Iwase	Tennoji Junior and Senior High School attached to Osaka Kyoiku University
Akeshi Kamae	Former Minami Junior High School, Osaka City
Akio Kawauchi	Osaka City University
Hirokazu Okamori	Emeritus Professor of Osaka Kyoiku University
Mituo Okuyama	Baika Junior High School, Osaka City
Hiroji Shibamoto	Hamadera Junior High School, Sakai City
Bunji Terada	Former Fukuizumi Junior High School, Sakai City
Mikiharu Terada	Shitennoji University
Hideo Tsuchida	Shitennoji High School
Tomoko Yanagimoto	Osaka Kyoiku University

6. Education Practices in Senior High Schools

In 6.1 and 6.2, prospects for students' capacity to understand knots invariants are discussed, which are based on experimental researches with mainly regular course students of private senior high schools. Their relevant background knowledges are also surveyed. In 6.3 and 6.4, experiments on knot invariants taken in a class of a senior high school attached to Osaka Kyoiku University are reported. In 6.5, a result that investigated whether knot theory is appropriate as a teaching material of the science course in Tennoji Senior High School, a prefectural senior high school, is reported.

6.1 A background of teaching knot theory in senior high school

6.1.1 A circumstance of senior high school in Japan   Students usually are familiar with making various forms of knots through their everyday life. In their childhood they used to play *ayatori* (cat's cradle), arranging a closed string into many different shapes. Today they use certain knots to tie things fixed from a shoelace to a necktie. Knot theory mainly asks how one can determine whether a given knot is a trivial knot or a nontrivial knot, whether two certain knots are equivalent or not, or whether they are deformed into each other. By assuming that an elementary school student is even capable of handling a theme of knots, our objective of this research is to measure a prospect of knots as a teaching material in the senior high school by investigating which types of knots a senior high school student can handle, how he or she can form an idea and deliver an abstraction on a knot. Or would he or she be able to transform his or her abstracted idea into an expression? In this chapter, three experiments at senior high schools are reported. In the first study by H. Kanaya at Seifu Senior High School, Osaka, the students experienced "*AWABI MUSUBI*" or a decorative knot used for gift-wrapping, "linking number" and "the bracket polynomial." They also explored a trendy topic on the latest DNA studies, which is actually related to a concept of knots. In the second study by N. Masuda at Kansai Soka Senior High School, the students employed "rank" as a knot invariant as their exercises. In the third study by M. Matsumoto at Osaka Gakuin Daigaku Senior High School, the students handled the bracket polynomial and Reidemeister move II. They ranged from junior high school students to senior high school third-year students and not all of them had shown a great interest in mathematics. These experiments are served to see the following points:

(1) What mathematical subjects are interested in the high school students who tend to feel stuck with the school mathematics.

(2) How they could create mathematics from ordinary items in everyday life.

(3) Which parts of the exercises would be interesting to most senior high schools uniformly taught required subjects for mathematics up to Mathematics I.

Actually, however, there is a wide gap in how far they get to learn mathematics depending on the courses they enroll. Although science students learn all of Mathematics II, III, A, B and C, besides Mathematics I as required, humanities students learn only up to Mathematics I, or just learn Basic Mathematics. When it comes to the area of geometry, they usually finish learning the subject of geometric demonstration in junior high school. They handle only figures in limited fields: Analytical geometry, Complex plane and Trigonometric ratio. The revised curriculum guidelines used from 2013 explain a content on "Applications of Mathematics", made by replacing the current Mathematics C and Mathematical Basics with the following mathematics:

(a) **Mathematics and human activities:** Students are to know relations between mathematics and cultures as well as to learn the fact that mathematics is originated in and developed along human activities.

(b) **Quantity and figure in human activities:** Students are to understand the involvement of mathematical concepts, e.g. numerical quantity, figure etc., in human activities and cultures.

(c) **Mathematics in plays:** Students are to recognize merits on logical thinking through playing mathematical games and puzzles and to know involvements of mathematics in cultures.

(d) **Social activity and mathematics:** Students are to review some mathematical situations in social activities through applying mathematics to relevant topics.

(e) **Devices for mathematical expressions:** Students are to mathematically review certain events, which are to be expressed in mathematical forms using diagrams, charts, matrix and discrete graphs.

(f) **Data analysis:** In order to deliver a forecast or a judgment based on tendency among the data, students are to garner relevant data according to their objectives and to process them by using spreadsheet, etc.

The revised curriculum guidelines used from 2013 are stated as follows:

Mathematical Content on *Course of Study* 2013 by *Ministry of Education, Culture, Sports, Science and Technology (MEXT)*:

• Mathematics I: Equations, Quadratic function, Figure and Measuring, Analyze of Data
• Mathematics II: Formula, Figure and Equation, Exponential function and Logarithmic function, Trigonometric function, Introduction of differentiation and integral
• Mathematics III: Curve in plane and complex plane, Limit, Differentiation, Integration

- Mathematics A: Combination, Permutation and probability, Property of integers, Property of figures
- Mathematics B : Probability distribution and statistical inference, Sequence, Vector

The mathematics curriculum of senior high school had been designed to lead students toward a goal of infinitesimal calculus under a great influence of J. Perry's reform in mathematics education in the early 1900's. Today, "an extended reproduction of math haters" is concerned about mathematics in senior high school. Some humanities students forget what they learned in math classes until the time of graduation, because they usually finish their learning math in their first year. What remedies can math teachers provide when they find students who lost interest in math at an early stage? How about bringing them back to where they had started learning math and re-starting it over?   Would math teachers be able to reintroduce students to modern mathematics from a different viewpoint? In this time, the class content is needed to be interesting and to keep in touch with everyday life. Would math teachers be able to provide students with chances of creating math on their own abilities or of touching modern mathematics? In order to make a breakthrough in the dilemma of math education, the area of discrete mathematics as a material to an experiment with students is taken up. Fortunately, this approach seems to have a better prospect for a success under the newly introduced curriculum on "Application of Mathematics". In the occasion of student workshops entitled "Fractal Theory" and "Fuzzy Logic", we surveyed the levels of their comprehension abilities in those areas. The new curriculum guidelines specify an assignment of Geometry teaching in senior high school as follows: In "Figure and Metrics" of Mathematics I, students learn Figure Using Trigonometric Ratio. In Mathematics II, students learn Figure and Equation; In Mathematics III, students learn Geometry Using Complex Plane. In Mathematics A, students learn Property of Complex Plane and Property of Figure including Property of Triangle and Circle, Construction and Solid Figure.   In Mathematics B, students learn Development on Vector. Questions coming to mind under this situation are how students would perceive the three-dimension space in which they actually spend their lives, when they should learn areas of Geometry relating to perception of the three-dimension space, and what sorts of perception they should be led to. Math teachers need to consider these points. We think that an introduction on trendy topics of knot theory to teaching in schools may make one of solutions.

6.1.2   Significance of teaching "Mathematical knots" in senior high school   We conceive that teaching "mathematical knots" in school may have significant merits in the following 5 points:

(a) **Fostering students' abilities of a spatial cognition by handlings of mathematical knots:** The conventional curriculum assigns Teaching of Descriptive Geometry in senior high schools as follows: Students learn "Figure and Quantity" with a focus on Trigonometric Ratio in Mathematics I; they learn "Figure and Equation" with focuses on Point and Line, Circle and Coordinate in Mathematics II. In Mathematics A, students handle Planar Figure containing Properties of Triangle and Quadrilateral, and Property of Circle. In Mathematics B, students handle Vector containing Vector in Plane and Vector with Space Coordinate. Although some opportunities of handling Figures in their classrooms are given, the handlings are limited in most cases to making exercises on Demonstration and Calculation. In other words, students have only few chances of learning the three-dimensional space. This is a reason why math teachers need to encourage students to learn Figures. One remarkable feature on "mathematical knots" experimenting as a new teaching material is that actual knots exist in the three-dimensional space although they are taught as figures called *diagrams* obtained by projecting onto the plane in a course of study. Thus, students handling this material are asked to transform a solid figure in the three-dimensional space into a diagram in the plane and conversely to transform a diagram in the plane into a solid figure in the three-dimensional space.   You can expect that these considerations would help students in fostering their abilities on a spatial cognition. In addition, handling a knot is expected to increase students' abilities to feel the space because students are to study knots by investigating an over-arc or an under-arc at every crossing point of the diagram.

(b) **Creating a "self-contained mathematics" by their own abilities:**   It is hard to deny that today's school mathematics or mathematics taught in a senior high school presents a passive learning experience, because, by working on math exercises, students earn little by little what is systematically prepared. At the same time, it deprives opportunities of students with positive attitudes to work out mathematics on their own abilities. "Mathematical knots" may propose a solution to this dilemma in school math. Students can experience creating a self-contained mathematics with a little ingenuity of their own abilities. With a little assistance of their teacher they can experiment taking a coloring on a *knot diagram* and compute the linking number by assigning an orientation to a link.

(c) **Giving an opportunity for students to learn a concept of invariants:** "Mathematical knots" offers an opportunity for handling a concept of topological invariants for students to examine whether two knots are substantially the same or not. Under the conventional curriculum in Japan, school mathematics provides little opportunity for handling a concept of invariants as a main theme. Exercises to compare two knots or links by computing a number invariant or a polynomial invariant for them contribute to fostering students' mathematical thinking and perspectives. "Mathematical knots" primarily request students to judge whether two knots or links are identical or not.   Then the students will fall into a concept of invariants.

(d) **Prioritizing mathematical thinking:**   If we realize that the conventional math curriculum tends to center on calculation exercises, the conventional evaluations on mathematical ability of students might have been inaccurate. In other words, it may be too simplistic to estimate that a student who is merely good at memorizing or handling formulae is good at mathematics. On the other hand, students can concentrate on a mathematical thinking by choosing a suitable topic on "mathematical knots".

(e) **Exhibiting significances of learning mathematics:**   The conventional math curriculum in Japan intends to lead students toward an acquirement of Infinitesimal Calculus.   It takes long time till students accept a reason or a merit why they should learn mathematics. On the other hand, "mathematical knots" can show students a clearer reason and a merit why they should learn mathematics, because not only it is now actively studied as one of the cutting-edge mathematics, but also it contributes now to various fields in science and technology relating to physics, chemistry and biology.

6.2 Education practices in private senior high school

6.2.1 How to discuss   Students attending the class were encouraged to consider the fact that there are various ways of tying strings for particular uses (For example, look for a different property between the knots in Fig.6.2.1).

Square knot   *Surgeon's knot*

Fig. 6.2.1

They examined how strings are formed in a necktie or how Kendo protective gears are fixed to body. *AYATORI game,* a play to form various figures of a trivial loop in their childhood, was also mentioned as an example of mathematical knots.   The fact that a use of knot patterns dates back until Jomon period, an ancient time in Japan, was also mentioned. Discussion between teacher and students in Seifu Senior High School:

T: How would you distinguish *CHOCHO MUSUBI* from *MARU MUSUBI* ?

CHOCHO MUSUBI   *MARU MUSUBI*   *HACHINOJI MUSUBI*

S1: *CHOCHO MUSUBI* can be easily untied while *MARU MUSUBI* cannot.

T: Let us think why some knots can be untied but others cannot. Let us start with a drawing of these knots. Can they be drawn on a paper? [1] In drawing a knot, let us pay attention to their crossing points.

S2: Since one strand is upper than the other strand, a crossing arises from an over-strand and an under-strand.

T: How about with *MARU MUSUBI* ?

S3: Every crossing is made of an over-strand and an under-strand as well.

T: I handed each of you out a colored string to try to make it.

S4: A colored string is got tangled easily.

T: What kinds of knots can you see in your everyday life? How many tying things can you say?

S5: *A* knot for fishing, *HACHINOJI MUSUBI*(a figure-eight knot). Another example is a knot used when one wears a protective mask for *Kendo*. The knot for fishing is a knot hard to untie it, but the knot for the protective mask needs to be easily untied when it is taken off.

T: Let us draw diagrams on a paper to compare a trivial knot and a nontrivial knot.

S6: How about counting the number of parts coming over the other and the number of parts coming under the other on a crossing point?

S7: The number of the over-strands and the number of the under-strands are equal in the case of a trivial knot. These numbers are not the same in the case of a nontrivial knot?

Students named examples of tying strings in everyday life: namely, the tying of a fishing line which cannot be untied as an example of a nontrivial knot and the tying of a Kendo protective mask which can be untied when finished as an example of a trivial knot. They examined these cases by using diagrams projected into a paper, although they could tell a certain type of knots which can be untied.

---

[1] Drawing a knot on a paper needs an instruction: a lower strand around every crossing point of a knot diagram should be drawn with a break.

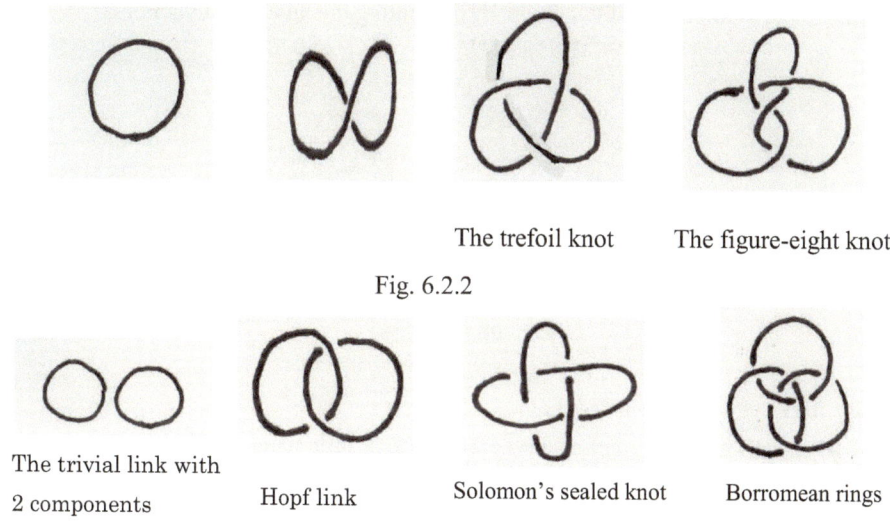

The trefoil knot   The figure-eight knot

Fig. 6.2.2

The trivial link with
2 components   Hopf link   Solomon's sealed knot   Borromean rings

Fig. 6.2.3

6.2.2 Education practice in Seifu Senior High School   We report here a practice on the bracket polynomial of "*AWABI MUSUBI*"(an abalone knot) (Photo 6.2.1) which is done in the electrical club in Seifu Senior High School in December 2008. Target students were two 2nd graders who belong to the electrical club. Three times guidance based on a text was done for them on untying a knot and computing a polynomial invariant to distinguish some knots. This practice was carried out as a club activity.

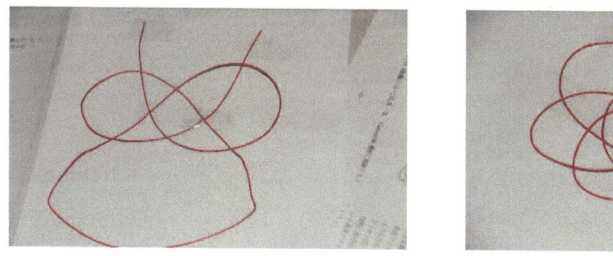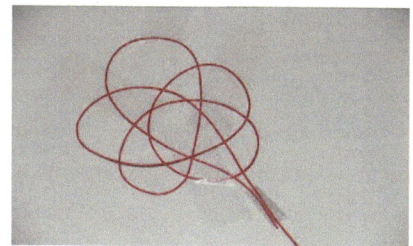

Photo 6.2.1

Teaching plan   The practices for 2 hours are as follows:

(a) To make an "*AWABI MUSUBI*".

(b) To use three strings to make a "*Misanga*" (a 3-braid) whose closure is the figure -eight knot. (Photo 6.2.2)

(c) Coloring around each crossing of a trefoil knot diagram.

(d)To make a decision whether or not a given knot and the mirror image has the same knot.

(f) To explain a difference of two knots by computing the bracket polynomials.

Photo 6.2.2

Senior high school student feedback

   - A knot was fun. A part of a separation was a little hard to understand. I enjoyed involving a few places.

   - I felt it amazing to think a "tie" mathematically.

6.2.3 Education practice in Kansai Soka Senior High School

In the practice here, we experimented with one 1st grade student, three 2nd grade students and one 3rd grade student of Kansai Soka Senior High School for 5 school hours in total.

Educational significances of teaching knot theory in senior high school

   (a) It is a good chance for students to understand a concept of invariants.

   (b) Students' experiences to make and do Mathematics by themselves.

   (c) Students can concentrate on a mathematical thinking without complicated calculations.

   (d) Students' senses of space are improved.

   (e) Students can clearly understand a significance of learning mathematics.

Target students and a placement of teaching knot theory in a curriculum   Our target students are the 1st grade students of the senior high school and, in the curriculum, we placed the unit "Teaching Knot Theory" after "Logic and Set" of Mathematics A. By counting the students' developmental stage, we decided to choose a *knot invariant* called "*rank*" as a topic which they can develop their ideas by using a numerical formula.   The objective of teaching this unit is to prove that we cannot untie the trefoil knot by employing a numerical formula.

Expected educational significances   By the rank which is a knot invariant, it is expected that we can achieve the educational significance 1 stated above. Also, through proving that the trefoil knot cannot be untied by using an equation, it is expected that we achieve the educational significances 2, 3 and 5. Furthermore, after we draw diagrams of actual knots and consider them, we can make a knot from any other diagram. Thus, we can also expect that the students achieve the educational significance 4.

Teaching plans on arrangements of the materials

1st class: Knots that can be untied and knots that cannot be untied.

2nd class: The minimum crossing number of a knot that cannot be untied.

3rd class: The trefoil knot and invariants.

4th class: Modular equalities and rank.

5th class: Solving equations by the modular equalities to prove that we cannot untie the trefoil knot.

Details of teaching plans on arrangements of the materials are introduced below.

1st class: Knots that can be untied and knots that cannot be untied

Introduction: Using the computer file "Power Point", we show various examples of knots, e. g., Ties, *OBI*, *MIZUHIKI*, and ribbons for a gift wrapping. Show shoes with a shoelace and introduce string knots such as a bow and a square knot.

Development: Showing a handout with a bow and a square knot illustrated, we make sure that we can untie any knot when we use an open string. Thus, we define a knot as follows: A knot is made by tying both ends of a string. Using a white string, we show that we can make a ring shape with a bow but we cannot make a ring shape with a square knot. Let the students consider a process of changing diagrams of a trivial knot into the circle. Let them teach moves in this process called *Reidemeister moves*. Make sure that they understand a trivial knot that can be untied and a non-trivial knot that cannot be untied. Also, let them realize that a trivial knot is a knot that can be deformed into a knot without crossing by using the Reidemeister moves. Let them understand that a non-trivial knot with one crossing number does not exist. Let them understand that a non-trivial knot with two crossing numbers does not exist.

2nd class: The minimum crossing number of knot that cannot be untied

Introduction: As a review of the last class, let the students make sure that non-trivial knots with one and two crossing numbers do not exist.

Development: Let them draw eight diagrams of knots with three crossing numbers. Let them image whether each diagram of knot can be untied or not. Make sure that they notice that six can be untied and two cannot be untied.

3rd class: The trefoil knot and invariants

Introduction: Return the handouts of the last class to the students and review two non-trivial knot diagrams.

Development: Let them put strings on the two diagrams to make knots. They can use a tape to keep the shape. After holding the both ends of the knots with a tape, take the knots from the paper. We make sure that these two knots cannot be untied. Teach some concepts on equivalence and invariants. Let the students consider what kind of a knot invariant exists.

4th class: Modular equality and rank

Introduction: We show that the set of integers is divided into the congruence classes on the reminders of integers divided by a fixed integer. Then introduce a modular equality.

Development: In a calculation of the modulus of an integer p, let the students complete the tables of additions and multiplications when $p = 3$, 4 and 5. Also, let them inform the concepts of complements and reciprocals. To define the rank, let them think of an arc in a knot diagram, a weight of the arc, a diagram with weights, a condition on the crossing point, an admissible weight, and a diagram with admissible weights. Let them make sure that there are several cases of admissible weights for a given knot diagram. We define this number of cases as the rank. Lead them to understand "If there is a knot whose rank is not zero modulo p, then the knot cannot be untied", because the rank modulo p of a trivial knot is zero.

5thclass: Solving equations using modular equality, giving the proof that we cannot untie the trefoil knot

Introduction: To teach how to solve an equation modulo p, we solve a linear equation when $p = 3$.

Development: Let the students solve the three questions on quadratic simultaneous equations, where there are three cases that there is one answer, there are more than one possible answers and the answer does not exist. Let them solve a system of three congruent linear equations modulo $p = 5$ in three unknown variables. Let them prove that the trefoil knot cannot be untied.

Feedbacks

1st class: Knots that can be untied and knots that cannot be untied

Introduction: We could introduce the topic smoothly by showing pictures, diagrams, actual shoes with a shoelace, leading them to be interested in the topic.

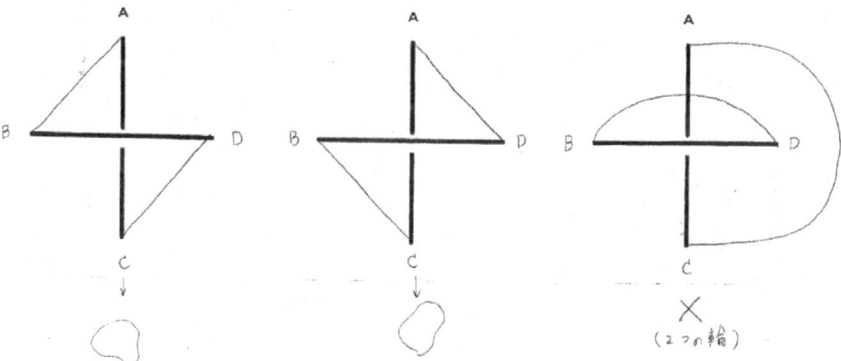

Development: Students naturally understood how they can draw a diagram from an actual knot by seeing the diagrams of the handout. When we ask them to consider a process of untying a knot, they naturally came up with Reidemeister moves I and II.

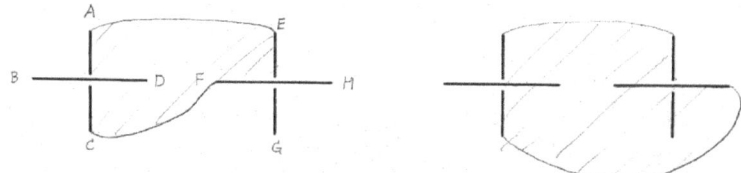

Fig. 6.2.4: Are there knots with one crossing that cannot be untied?

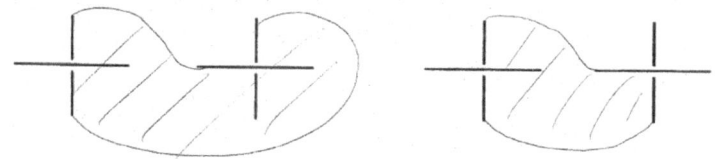

Fig. 6.2.5: Are there knots with two crossings that cannot be untied?

However, they could not think of Reidemeister move III. We needed to add an explanation of Reidemeister move III. It was easy for them to understand that non-trivial knots with one crossing do not exist.

From students' handouts: It was also smoothly understood by students that non-trivial knots with two crossings do not exist.

2nd class: The minimum crossing number of knot that cannot be untied

Introduction: Since the second class was held four months later due to the school schedule, we carefully review the main points of the last class, namely non-trivial knots with one or two crossings do not exist.

Development: Regarding that it is difficult for them to draw knot diagrams with three crossings, we give them a hint. As the result, we can save the time spending to draw and all of them could draw the eight diagrams correctly. Then we asked them to imagine which diagrams are trivial knots. All the students except one student could reach the correct answer. The hint was to explain that the first two lines should be drawn as follows:

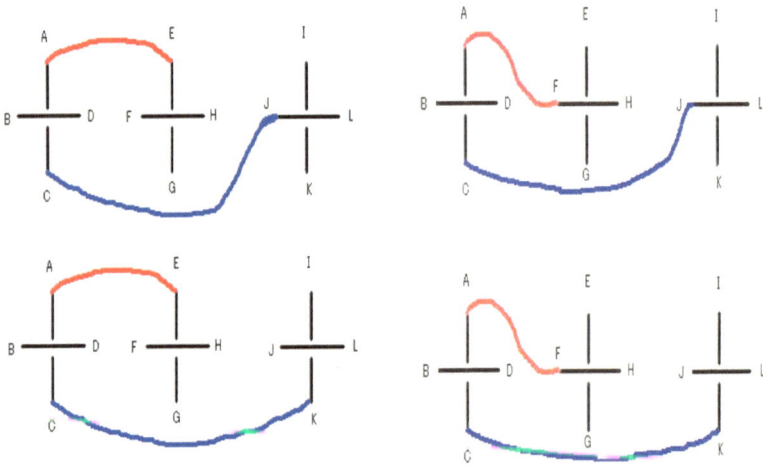

Fig. 6.2.6

Photo 6.2.3: Drawing the knots with three crossing numbers.

Photo 6.2.4: Drawing the knots with three crossing numbers

From students' handout: Diagrams with ○ are the knots that cannot be untied.

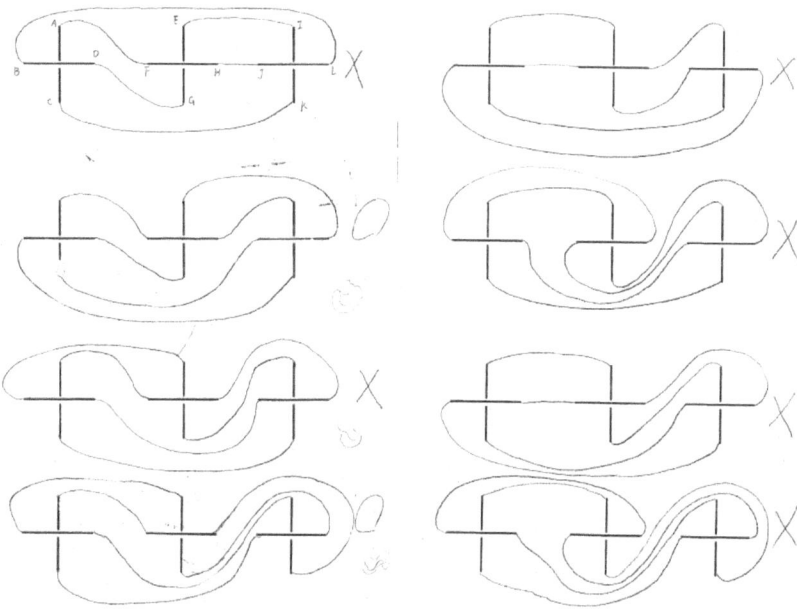

Fig. 6.2.7

Feedback from a student: *It was difficult to image which knot can be untied or not, but it was at the same time enjoyable to solve this question using the theory. (I understand that he meant he used his brain.)*

3rd class: The trefoil knot and invariants

Development: To make sure what they imagined in their mind is correct, we had the students make actual knots with strings on the two diagrams which, they judged, cannot be untied. They held the shapes of the knots by a tape.   We alerted them to be careful when they are working on the crossing points. Let them take off the knots from the paper and ask them to make sure that the knots cannot be untied. Also, let them move the knots and make sure that they are the square knots and at the same time the trefoil knots. Since unexpectedly it took time to make a knot with a string, we had to teach the concepts of equivalence and invariants quickly. Although there are slightly differences in explanation, all of the students understood an invariant to determine which knot can be untied or not.

Knot invariants: Tracing the strings, some students realized that a point going along the string has to go up or down alternatively around the crossing points. However, this is neither a necessary condition nor a sufficient condition for a knot to be non-trivial. In fact, in this diagram (Fig. 6.2.8), a point going along the string go through up and down at the crossing points but it can be untied.

Fig. 6.2.8

Thus, we informed them that we will think of an invariant using a numerical formula in the next class.

Photo 6.2.5: Making knots
with strings.

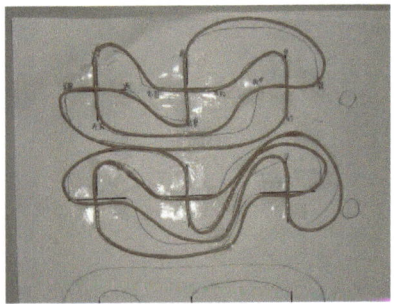

Photo 6.2.6: The knots with three
crossing numbers that cannot be untied

Photo 6.2.7: Holding the knots
that he made

Photo 6.2.8: Holding the knots
that he made

4th class: Modular equality and rank

Introduction: At first, explain by example that the set of integers is divided into the groups (congruence classes) on the reminders of integers divided by a fixed integer. The following table shows that 7 groups were made using the reminders of the integers divided by 7:

Group7	Group1	Group2	Group3	Group4	Group5	Group6
	1	2	3	4	5	6
7	8	9	10	11	12	13
14	15	16	17	...	...	...	...	...
...	...	...	...	...	...	...	...	...	...	...

Then we introduced the definition of the congruence and the symbol $\equiv$.

Development: Let the students consider the calculations of additions and multiplications with modulo p when $p=3$ and $p=5$. When they filled the answers in the chart, there are some students who misunderstood that $2+2\equiv4$ when $p=3$ and $3+4\equiv7$ when $p=5$. Therefore, we gave those students a detailed instruction and improved their understanding. When we gave the students an optional question with the even number $p=4$, all the students including the ones who misunderstood could answer correctly. The students shared their feedbacks and the class was a very mathematical class since they used numerical formulas. All the students could calculate a numerical condition on the crossing points needed to define the rank, but they asked us the reason why we should think of the condition.

5th class: Solving equations using modular equality and proof that we cannot untie trefoil knots

Development: Since it was the first time for the students to solve modular equations, we made handouts with overall explanations of how to solve these questions. These handouts also include some blanks where they can fill in. The handouts were very effective for them to understand the main points.

Feedbacks from the students: *"First of all, I didn't understand how a knot of a string is related to the mathematic, but learning at this class: 5th class, I understood the relation and enjoyed solving equations modulo 2 and 3 rather than solving usual equations. ", "Since I did not quite understand the handout of 4th class I want to review it. If I understand the point, I could perfectly master the main points of this unit! It was a great opportunity to learn a new numerical formula using congruence. I thought that it would be more difficult to solve them without a hint. I would like to prove that a knot can be untied by a numerical formula."*

Thought   In the 1st class, we considered whether certain knots can be untied or not by changing actual knots into knot diagrams.   In the 2nd class, we found out that the minimum crossing number of non-trivial knots is three by using the knot diagrams.   In the 3rd class, we made actual knots from the knot diagrams with three crossings. Through instructions of knot theory, we experienced a transforming a three-dimensional figure into a two-dimensional diagram, considering moving two-dimensional diagrams and transforming a two-dimensional diagram into a three-dimensional figure.   As the result, we could improve students' sense of space, which is the educational significance (d) in 6.2.3.1. In addition, through learning the concepts on the rank which is a knot invariant using modular equalities, we could prove that trefoil knots cannot be untied and in conclusion we could achieve the following educational significances: (a) It is a good chance for students to understand a concept of invariants.   (b) Students experience to make and do mathematics by themselves.   (c)Students can concentrate on a mathematical thinking without complicated calculations. (e) Students can clearly understand a significance of learning mathematics. (f) Experiencing this educational experiment and considering a developmental stage of high school students, we found that students were fascinated by the fact that they could judge whether a knot can be untied or not by using a numerical invariant.

As a next task, we would like to develop a teaching material of polynomial invariants such as Jones polynomial and to use it for an actual class.

6.2.4   Education practice in Osaka Gakuin Daigaku Senior High School   This research was made for 3 students in the 2nd and 3rd grades of Osaka Gakuin Daigaku Senior High School on December 2008 in total 6 school hours. These students are council officers and think they are not so good at math

The purpose of research   The purpose of the research is as follows:
- We investigate whether high school students can understand easier a state that is loops in the plane without crossings when they learn the bracket polynomial.
- We investigate whether they can voluntarily derive a formula whose rule is simple.

Teaching Plan
    (1) What is a knot?
    (2) Reidemeister moves and studying the states obtained by splicing every crossing of a knot diagram.
    (3) The bracket polynomial.
    (4) The bracket polynomial of a knot with many crossings.
    (5) Relationship between the bracket polynomials of a knot and the mirror image.

Details of the contents in teaching plan introduced below.

1ˢᵗ class: What is a knot ?

Introduction: Let the students make a knot. Teacher explains the definition of a mathematical knot. Teacher explains the edges of the string used in tying.

Expanding the teaching development: Teacher has the students make a knot (with up to three crossings). Teacher has them think whether or not the knot can be untied and then think to make an inextricable knot. Teacher explains to them a trivial knot and a typical non-trivial knot such as the trefoil knot.

Examples of questions:

Q1-1: Make knots with one, two and three crossings.

Q1-2: By examining a knot, try to manipulate knots similar to the knots.

Q1-3: Make the trefoil knot and a trivial knot and then try to make the knots occurring from putting them in the mirror, i.e., the *mirror images* of them.

Impression of class: Almost all students understood that a non-trivial knot has at least one crossing. However, for some students it was needed to explain that a non-trivial knot has at least two or three crossings. It was hard to produce a non-trivial knot. Students realized that the knot seen in a mirror is the same as the knot seen in the reverse side of the tracing paper. The picture in Photo 6.2.9 was taken at that time.

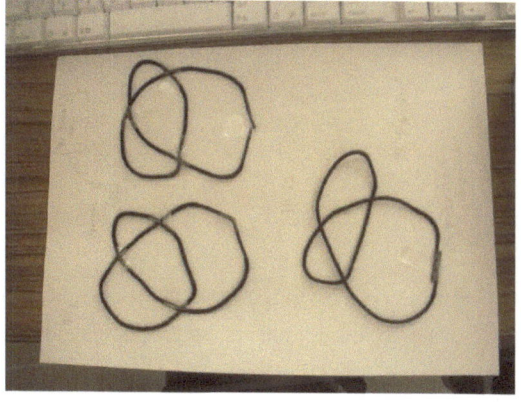

Photo 6.2.9

2ⁿᵈ class: Reidemeister moves and study of the states obtained by splicing the crossings of a knot diagram

Introduction: The students review the definition of a knot and a typical non-trivial knot.

Expanding the teaching development: Teacher listens to their opinions on how to deform a string. Teacher teaches the Reidemeister moves and a state obtained from a knot diagram. Let the students make an analysis on states of knots with one or two crossings.

Examples of questions:

Q2-1: Illustrate the A-splice and the B-splice by giving a crossing.

Q2-2: Try to make a state with two loops from a knot diagram with one crossing.

Q2-3: Try to make a state with two loops from a knot diagram with two crossings, and then try further to make a state of four loops.

Q2-4: Try to make a state with 8 loops from a trefoil knot diagram with three crossings.

Q2-5: How many loops are there in a state obtained from a knot diagram with n crossings?

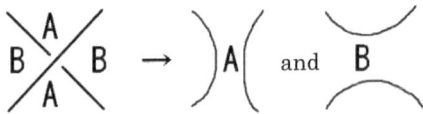

Fig. 6.2.9: A-splice and B-splice

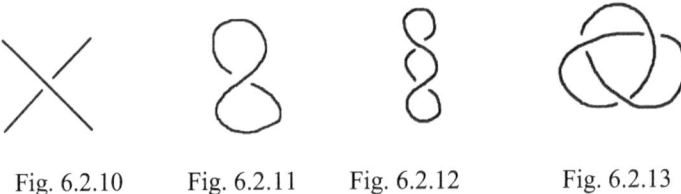

Fig. 6.2.10   Fig. 6.2.11   Fig. 6.2.12   Fig. 6.2.13

Impression of class: As previous research results, the students soon noticed the Reidemeister move I. Only some students noticed the Reidemeister move II. Nobody noticed the Reidemeister move III. When some students operated the Reidemeister move II to eliminate two crossings, they pulled the ends of the string. Only some students could understand a state soon. Many students could not obtain a state with two loops from a knot diagram with two crossings in Q2-3. However, they could obtain a state from a knot diagram with three crossings easily. The lessons were continued and, in fact, many students could understand the states on a trefoil knot diagram for three hours.

3rd class: The bracket polynomial

Introduction: Students reviewed about how to splice a crossing of a knot diagram.

Expanding the teaching development : Teacher explained them the bracket polynomial. Next, they made an expression based on the number of loops obtained by splicing from a knot diagram, so that they could obtain a polynomial by summing all the expressions. Thus, they could calculate the *bracket polynomial* of a knot diagram with one crossing and then it of a knot diagram with two crossings.   After then, they tried to calculate the bracket polynomial of the trefoil knot.

Examples of questions:

Q3-1: Calculate the bracket polynomial from the states obtained from the knot diagram with one crossing in Q2-2.   (The answer: $Ad+Bd^2$)

Q3-2: Calculate the bracket polynomial from the states obtained from the knot diagram with two crossings in Q2-3.   (The answer: $Ad^2+B^2d^2+ABd^3+ABd$)

Q3-3: Calculate the bracket polynomial from the states on a trefoil knot diagram with three crossings.   (The answer: $3A^2Bd+(A^3+3AB^2)d^2+B^3d^3$)

Impression of class: According to the advancement condition of a study, the students came to have a difference in their learning level. After students with slow progresses finished the classification of two crossings, Teacher taught all the students how to compute the bracket polynomial. Although some students had poor senses of mathematics, they could compute the bracket polynomial of the trefoil knot.

 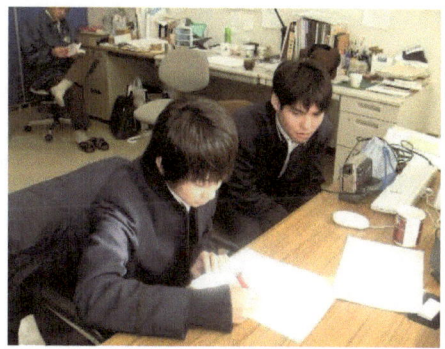

Photo 6.2.10   Photo 6.2.11

4th class: The bracket polynomial of a knot with many crossings. I

Introduction: The students obtained the bracket polynomial of the trefoil knot by classifying the states.

Expand Teaching Development: After computing the bracket polynomial of the trefoil knot , they tried to compute the bracket polynomials of knot diagrams with 3 and 4 crossings.

Example of questions:

Q4-1: Following an instruction, compute the bracket polynomial of a given knot diagram.

Q4-2: The knot diagram with two crossings of Q2-3 is a trivial knot by the Reidemeister move II. Then compare bracket polynomial of the knot diagram of Q2-3 with the bracket polynomial of a trivial knot diagram to find identical relationships between A and B, and between A and B.

Impression of class: In this time Teacher gave the student the problems on Q4-1.

Fig. 6.2.14

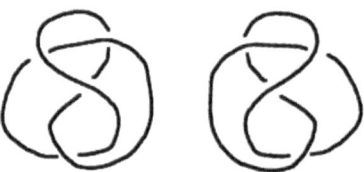

Fig. 6.2.15

Student A in the 2nd grade: A trivial knot diagram analogous to a trefoil knot diagram.
Student B in the 2nd grade: The mirror image of a trefoil knot diagram.
Student C and Student D in the 12th grade: A trivial knot diagram analogous to a figure-eight knot diagram and the mirror image.

One of the 3rd grade students solved so quickly an additional problem to find an identical relationship on the Reidemeister move II. At this time, it became an overtime of the schedule. They could not solve the relationship "$AB\,d = d$". After making the student contemplate it, Teacher made the following explanation in a consideration of the understanding of the students. Because some of students was not good at math, they hardly derived "$AB = 1$" from "$B = A^{-1}$". The 2nd grade students ended the class at this time. After examining the time distribution, Teacher thought that the students should have a time to examine the mirror image of a knot.

Answers to the problems on Q4-1:
The mirror image of the trefoil knot in the middle of Fig. 6.2.14:
$3AB^2d+(B^3+3A^2B)\,d^2 + A^3\,d^3$.
A trivial knot diagram analogous to a trefoil knot diagram in the right of Fig. 6.2.14:
$(A^3+2AB^2)\,d+(B^3+3A^2B)d^2+AB^2d^3$.
The trivial knot diagram analogous to the figure-eight knot in the left of Fig. 6.2.15:
$(B^4+4A^2B^2)d+4(A^3B+AB^3)d^2+(A^4+2A^2B^2)d^3$.
The mirror image of the trivial knot diagram analogous to a figure-eight knot diagram in the right of Fig. 6.2.15:
$(A^4+4A^2B^2)d+4(A^3B+AB^3)d^2+(B^4+2A^2B^2)d^3$.
Deriving an identical relationship on the Reidemeister move II, we have that the polynomial $A^2d^2+B^2d^2+ABd^3+ABd$ is identically equal to d, so that we have $ABd=d$ and $AB = 1$. Further, we have $A^2d^2+B^2d^2+ABd=0$ and $A^2 + ABd + B^2 = 0$.
Thus, we have $B = A^{-1}$ and $d = -A^2 -B^2 = -A^2 -A^{-2}$.

5th class: The bracket polynomial of a knot with many crossings. II)
 Introduction: Since a time passes from the last class, the students reviewed the bracket polynomial.
 Expand Teaching Development: The students continued to compute bracket polynomial of a trivial knot diagram analogous to a figure-eight knot diagram.
 Impression of class: One of the students succeeded in computing the bracket polynomial while he was seeing a figure of the states of a knot diagram. He is good at design. He could consider the figure in his head. The figure described below is what he has drawn.

 Another student solved a problem by drawing a computer software of Adobe Illustrator. It was easier to him than drawing by his hand. The figures in Fig. 6.2.17 and Fig. 6.2.18 are what has drawn by him.

116

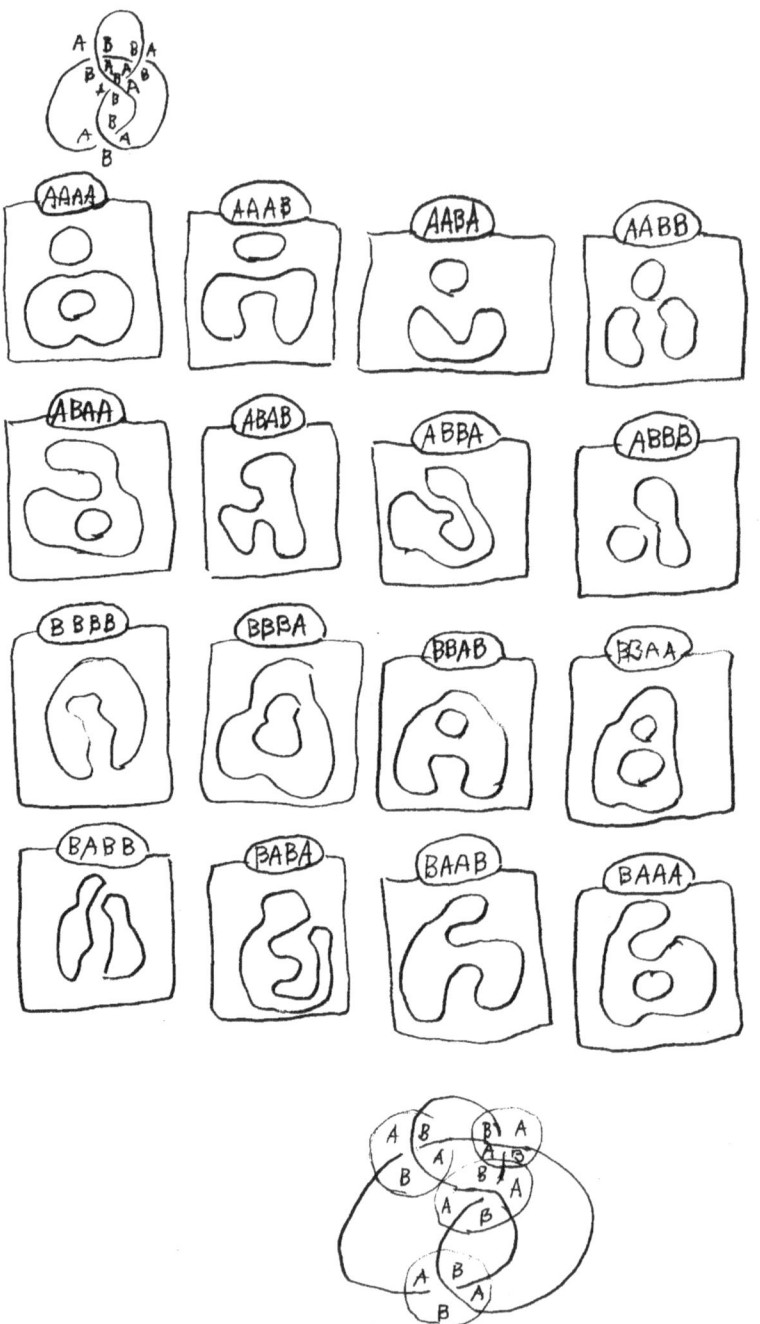

Fig. 6.2.16: The figure of *A*-splices and *B*-splices (by a student)

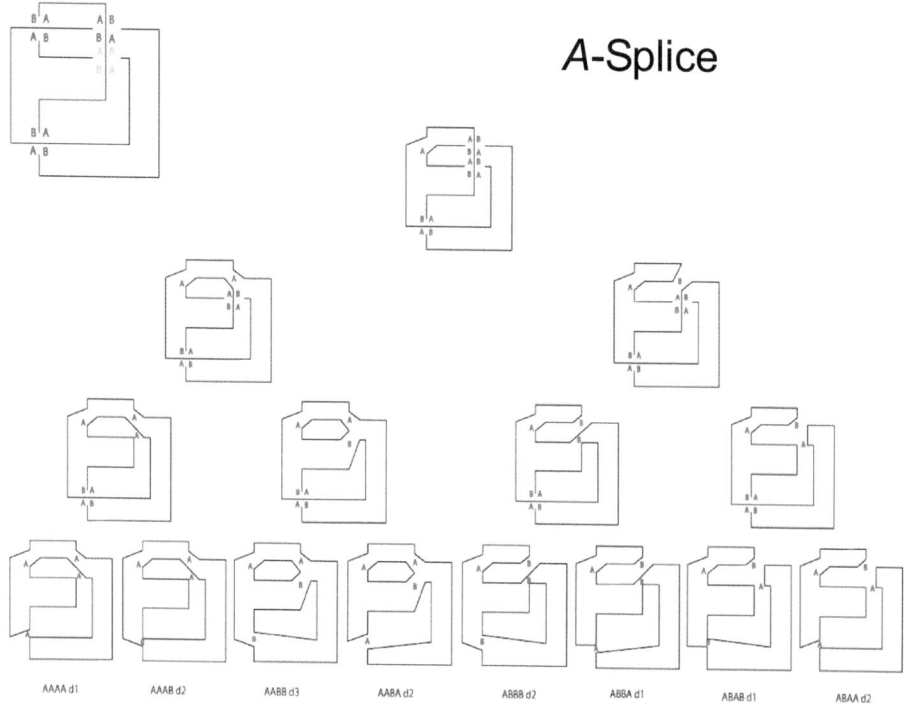

Fig. 6.2.17:   *A*-splice

6th class: Relationship between the bracket polynomials of a knot and the mirror image

Introduction: The students confirmed the bracket polynomials of the knots.

Expand Teaching Development: They thought about a relationship between the trefoil knot and its mirror image. Using the results, they confirmed bracket polynomial in the states of a trivial knot diagram analogous to a figure-eight knot diagram. They thought whether it is possible to obtain the same result in some other cases.

Examples of questions:

Q6-1: Compute the bracket polynomials of a trefoil knot diagram and its mirror image, and a trivial knot diagram. What relations can be found among these bracket polynomials?

Q6-2: Compute the bracket polynomials of a figure-eight knot diagram and its mirror image. Compare this results with the results in Q6-1. What kinds of differences are there between the results of Q6-1and Q6-2?

Q6-3: Does a relation found in Q6-1 hold for any other knot? If so, then why?

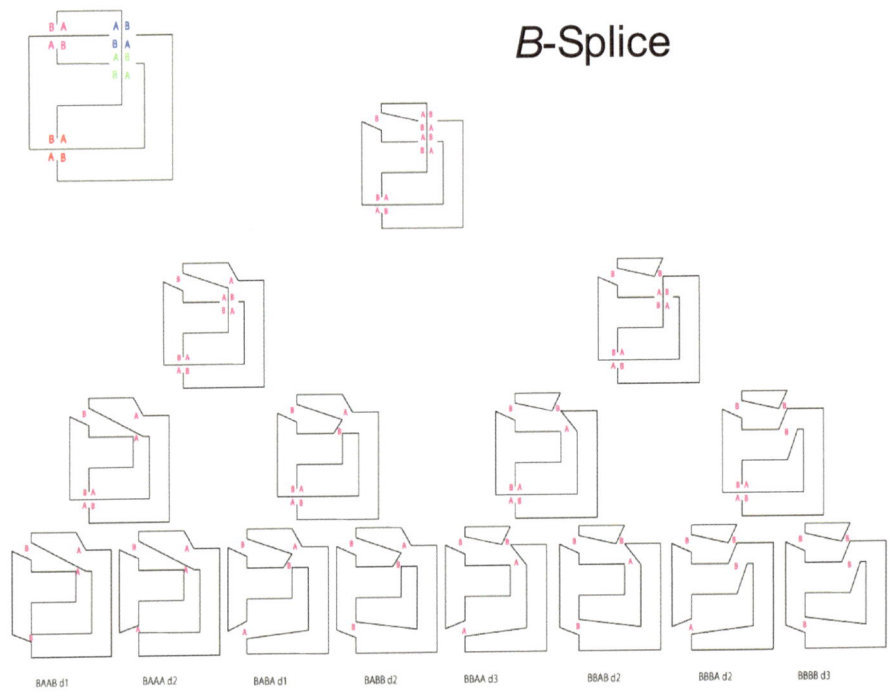

B-Splice

Fig. 6.2.18:   B-splice

Photo 6.2.12

Impression of class: Because it is necessary to compare three knots, the students thought Q6-1 for a little while. They noticed that the bracket polynomial of the mirror image is obtained from the original bracket polynomial by exchanging A into B. On Q6-2, when the relation obtained by Q6-1 is substituted, the result was not the case. However, when they calculated it again, they noticed that the calculation contained an error and the relation was correct. On Q6-3, the current experience was made in the best use and the student was deriving answers in the mirror image and a hierarchical relationship at the crossing.

Feedback from students

Student A: *It was very difficult. I knew that it was possible to make it to mathematics even by this simple event.*

Student B: *It was troublesome. Unlike a regular math class, though the definition is shown, I had to invent a formula by myself and to confirm it, so that I felt that my brain in the part not used was activated.*

Student C: *I had a feeling that had returned to the grade-schooler. I tried to think it deeply though I thought that it did too much. I recalled that I thought such a matter deeply in an old time.*

Consideration   In considering expressions of a knot by the bracket polynomial a teacher can give the students a lot of parts that they voluntarily think of, because the definition of the bracket polynomial is simple.   In particular, the students can understand it easily to derive the relations from the Reidemeister move II using the two crossings of a trivial knot, because it is possible to think an evident circle "d".   There were much time to investigate the Reidemeister moves I and III this time, more classes are hoped to have in future. The relation on an expression of the bracket polynomials of a knot and its mirror image is very simple if we understand that it is the exchange of A and B.   However, the students were continuing to calculate the expressions each other and to verify it each other until they noticed the relation.   Teacher felt that the students understood "Universality" for the first time through this process.   One student told his views to Teacher "I wonder whether various people thought 'I'm great!' or whether they thought 'Such a thing is natural' when they derived the law". Teacher felt he experienced deriving a law dim. Through this experiment, Teacher thought that learning mathematics of a knot is easy to do mathematical consideration especially in following sense: Students can catch things in visual even if his calculating ability is somewhat low. However, Teacher should also examine the opposite case in future. In a next step, Teacher needs to discover a new thing so that the students can practice problems on the other spatial awareness together when they practice mathematics of a knot.

120

References

[1] A. Kawauchi et al, Knot Theory (in Japanese), Springer Verlag, Tokyo, 1990. (English expanded version: A Survey of Knot Theory, Birkhäuser Verlag, 1996.)

[2] A. Kawauchi and T. Yanagimoto et al, An Approach to Teaching Knot Theory in School Mathematics for Pupils and Students (in Japanese), Project of Teaching Knot Theory in School Mathematics, Research Report as Educational Action in 21st Century COE Program "Constitution of wide-angle mathematical basis focused on Knots(Osaka City University)" Vol.1 2005, Vol.2 2007, Vol.3 2009.

6.3 Education practice in senior high school attached to Osaka Kyoiku University (1)

6.3.1 Introduction   In this section we show our practices in Tennoji Senior High School attached to Osaka Kyoiku University. We took up two teaching materials in Knot theory. One is a knot invariant, which we call X-polynomial and another is the mirror image of a knot. We were very interested in taking up these materials as teaching materials in high schools. We first introduce some contents of these materials.

The X-polynomial   To define this knot invariant, we need the following bracket polynomial and the concept of Reidemeister moves.

The bracket polynomial   For a link diagram L, we define the polynomial $<L>$, called the *bracket polynomial* of L, by the 3 rules described as below. For example, $<\text{⬤}>$ denotes the bracket polynomial of knot ⬤.

The *3 rules for the bracket polynomial* in unknown variables A,B,C

(i) $<\bigcirc>=1$.

(ii) $<\text{✕}> = A<\text{)(} >+B<\text{≍}>$, $<\text{✕}> = A<\text{≍}>+B<\text{)(} >$.

(iii) $<L\cup\bigcirc>=C<L>$.

For (ii), we understand that for the transformation $\text{✕}\rightarrow\text{)(}$, we put A as the coefficient of $<\text{)(} >$ and for the transformation $\text{✕} \rightarrow \text{≍}$, we put B as the coefficient of $<\text{≍}>$.   For (iii), we understand that when we remove the loop component $\bigcirc$ from the diagram $L\cup\bigcirc$, we put C as the coefficient of $<L>$. Here is an example.

Example $<\text{⬤}>=A<\text{⬤}>+B<\text{⬤}>$

$$=A\{A<\text{⬤}>+B<\text{⬤}>\}+B\{A<\text{⬤}>+B<\text{⬤}>\}$$
$$=A^2<\bigcirc\bigcirc>+AB<\bigcirc>+BA<\bigcirc>+B^2<\bigcirc\bigcirc>$$
$$=A^2C<\bigcirc>+AB<\bigcirc>+BA<\bigcirc>+B^2C<\bigcirc>$$
$$=A^2C+AB+BA+B^2C$$

Reidemeister moves   The *Reidemeister moves* are defined by the following three local moves:

122

Reidemeister move I	Reidemeister move II	Reidemeister move III

In 1926, K. Reidemeister, a German mathematician, observed that if we have two distinct diagrams of a knot, we can obtain from one to the other by a series of Reidemeister moves I, II, III and planar deformations. Although each of these moves changes a shape of the knot, it does not change the knot itself. To be that $<L>$ is a knot invariant, we need, for example, $<\bigcirc\!\!\!\!\!\bigcirc> = <\bigcirc\bigcirc>$.

Checkpoint 1: If we have $<\mathord{\gtrless}> = <\mathord{)(}>$ identically, then $B = A^{-1}$, $C = -A^2 - A^{-2}$.

Checkpoint 2: If we define B and C as in Checkpoint 1, then the Reidemeister move II does not changes the bracket polynomial.

Checkpoint 3: If we define B and C as in Checkpoint 1, then the Reidemeister move III does not change the bracket polynomial.

The *X-polynomial* $X(L)$ is defined by the bracket polynomial $<L>$ and the *writhe* defined as follows: Let's give an orientation to a link diagram L. We assign every *crossing* of L to $+1$ or -1 as follows: $\diagup\!\!\!\!\diagdown \Rightarrow +1$, $\diagup\!\!\!\!\diagdown \Rightarrow -1$
Then the writhe $w(L)$ is defined to be the sum of the sign ± 1 for all crossings of L. The *X-polynomial* $X(L)$ is defined by $X(L) = (-A^3)^{\cdot w(L)} <L>$.

For example, a computation of $X(L)$ is made as follows:

Example: $X(\bigcirc\!\!\!\!\!\bigcirc) = (-A^3)^{-w(\bigcirc\!\!\!\!\bigcirc)} <\bigcirc\!\!\!\!\!\bigcirc>$ for oriented link $\bigcirc\!\!\!\!\!\bigcirc$.

In fact, $w(\bigcirc\!\!\!\!\bigcirc) = +2$ and

$$<\bigcirc\!\!\!\!\!\bigcirc> = A^2C + AB + BA + B^2C$$

$$= A^2(-A^2-A^{-2}) + AA^{-1} + A^{-1}A + (A^{-1})^2(-A^2-A^{-2})$$

$$= -A^4 - 1 + 1 + 1 - 1 - A^{-4} = -A^4 - A^{-4},$$

so that $X(\bigcirc\!\!\!\!\bigcirc) = (-A^3)^{-2}(-A^4 - A^{-4}) = -A^{-2} - A^{-10}$.

The X-polynomial $X(L)$ is unaffected by the Reidemeister moves I , II , III .

The mirror image of a knots   The *mirror image* of a knot is obtained from the original knot by interchanging the upper arc and the lower arc in a neighborhood of every crossing point.

Example: The mirror image of is .

6.3.2 Our purposes   The following (1)-(4) were our purposes.
(1) Let our pupils be interested in the mirror image of a knot by using an actual string.
(2) Let our pupils understand how to make the X-polynomial of a knot.
(3) Let our pupils understand that the X-polynomial is useful in finding a difference between a knot and the mirror image.
(4) Let our pupils be interested in that we can make the X-polynomial of a knot.

Teaching plan   Our flowchart and our practice are stated as follows.
   (a) The 1st period: "Is the mirror image of the trefoil knot the same as the original knot?"
   (b) The 2nd period: "How is the bracket polynomial made?"
   (c) The 3rd period: "How is the X-polynomial made?"
Our practice

The 1st period:
(i) Ask the question: What is the mirror image of a knot? Here, let's show the pupils the trefoil knot and the mirror image, and draw their diagrams. ⇒   The

mirror image of is .

(ii) Mention that the mirror image of is . Then ask the question: Are they the same link? ⇒Yes! They are same links.

(iii) Mention that the mirror image of is .   Then ask the question: Are they the same knot? ⇒ It is difficult to see whether they are the same or not!

(iii-1) Look at this picture . This is called the *figure-eight knot*.

(iii-2) Let's make the *figure-eight knot* by using a string.

(iii-3) Let's draw a picture of the mirror image of .

(iii-4) Let's try to change to by using a string. ⇒

and are the same knot. It is very interesting!

(iii-5) Please explain a flowchart to change into . ⇒ Let's

draw pictures to show changing view to . The flowchart is

described as follows:

(vi) Ask the question: Is the same knot as ? ⇒ Let's try to change

to . ⇒ It is difficult to change into . ⇒ Ask the

question: Is not the same knot as ? We cannot understand whether

they are the same or not.

The 2nd period
(i) Let's make the bracket polynomial of a knot. ⇒ Mention that we can compute
the bracket polynomial of every knot in unknown variables A, B, C by using the
following 3 rules, and that the same knot will have the same polynomial!

① $<\bigcirc>=1$.

② $<\times>=A<\rangle\langle>+B<\asymp>$, $<\times>=A<\asymp>+B<\rangle\langle>$.

③ $<L\cup\bigcirc>=C<L>$, where L is a link diagram.

(i-1) Let's make a bracket polynomial of 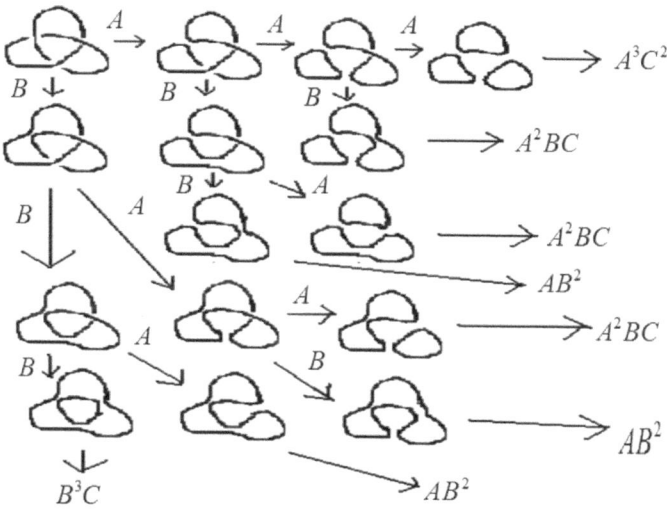. ⇒ Let's draw all the resulting pictures to record which rules you used for changing the diagram.

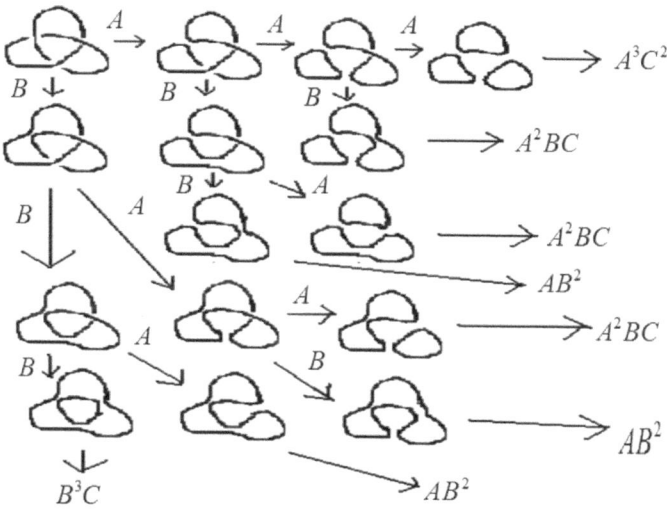

Then the bracket polynomial of 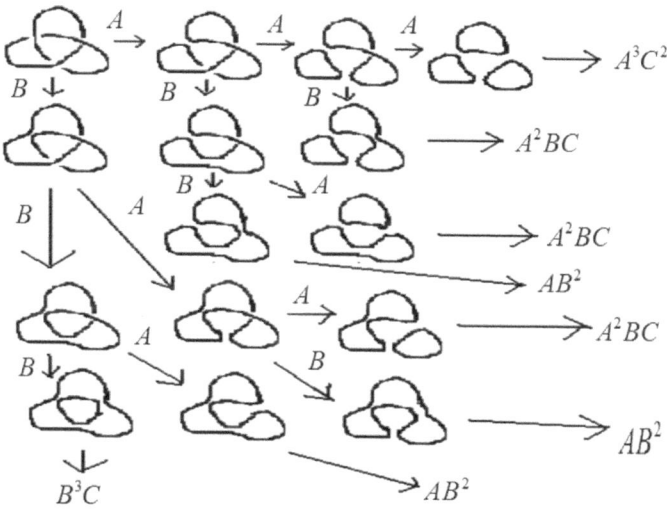 is

$$A^3C^2+3A^2BC+3AB^2+B^3C.$$

(i-2) Let's make the bracket polynomial of 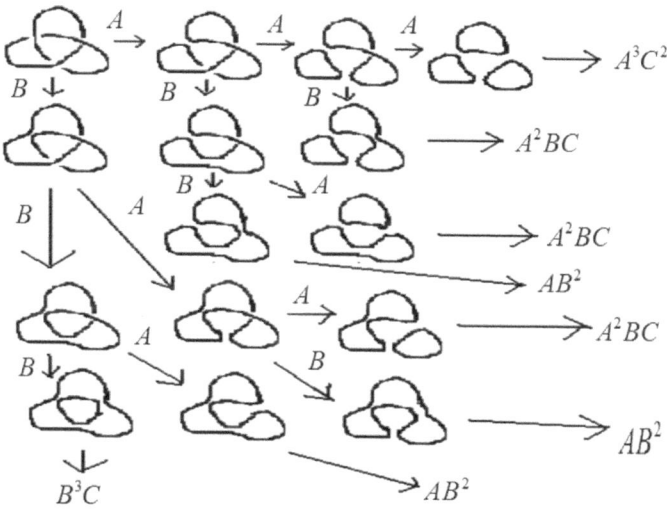. ⇒ We can obtain the bracket polynomial of this knot in a similar way. The bracket polynomial of 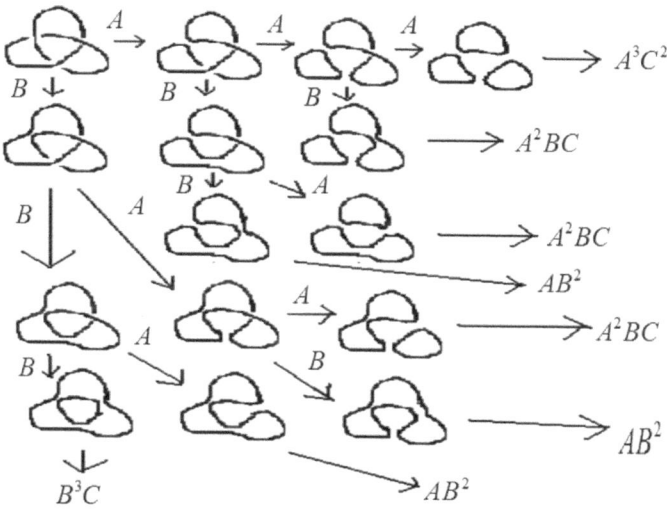 is $A^3C+3AB^2C+3A^2B+B^3C^2$.

(ii) Let's check whether the bracket polynomial changes or not by the Reidemeister moves described as follows:

I. ⟷   II. ⟷   III. ⟷

If we have two distinct diagrams of the same knot, we can obtain from one to the other by a series of Reidemeister moves and planar deformations. We think that it is necessary to teach the pupils "Reidemeister moves" previously.

(ii-1) Let's check whether the bracket polynomial of a knot do not change or not by Reidemeister move II. ⇒ Let's assume that the bracket polynomial is unchanged for a knot diagram and an another knot diagram obtained from it by Reidemeister move II. ⇒ We need ⟨ ⟩=⟨OO⟩. ⇒ Let's check

Reidemeister II by the 3 rules of the bracket polynomial.

$$\langle \text{⤫} \rangle = A\langle \text{⤫} \rangle + B\langle \text{⤫} \rangle$$

$$= A(A\langle \text{⤫} \rangle + B\langle \text{⤫} \rangle) + B(A\langle \text{⤫} \rangle + B\langle \text{⤫} \rangle)$$

$$= A(A\langle \text{)(} \rangle + B\langle \text{≍} \rangle) + B(AC\langle \text{)(} \rangle + B\langle \text{)(} \rangle)$$

$$= (A^2 + BAC + B^2)\langle \text{)(} \rangle + AB\langle \text{≍} \rangle = \langle \text{≍} \rangle$$

$\Rightarrow$ We must have $B = A^{-1}$ and $C = -A^2 - A^{-2}$. By assuming these identities, the three rules of the bracket polynomial are described as follows:

① $\langle \bigcirc \rangle = 1$.

② $\langle \text{⤬} \rangle = A\langle \text{)(} \rangle + A^{-1}\langle \text{≍} \rangle$, $\langle \text{⤬} \rangle = A\langle \text{≍} \rangle + A^{-1}\langle \text{)(} \rangle$.

③ $\langle L \cup \circ \rangle = (-A^2 - A^{-2})\langle L \rangle$, where L is a link diagram.

(ii-2) Let's check whether the bracket polynomial of a knot does not change or not by Reidemeister move III. $\Rightarrow$ We understand that the bracket polynomial of a knot does not change by Reidemeister move III.

(ii-3) Let's check whether the bracket polynomial of a knot does not change or not by Reidemeister move I.

$$\langle \text{⟲} \rangle = A\langle \text{⟲} \rangle + A^{-1}\langle \text{⟲} \rangle = A\langle - \rangle + A^{-1}(-A^2 - A^{-2})\langle - \rangle = -A^{-3}\langle - \rangle,$$

$$\langle \text{⟲} \rangle = A\langle \text{⟲} \rangle + A^{-1}\langle \text{⟲} \rangle = A(-A^2 - A^{-2})\langle - \rangle + A^{-1}\langle - \rangle = -A^3\langle - \rangle.$$

$\Rightarrow$ We understand that the bracket polynomial of a knot is changed by Reidemeister move I. $\Rightarrow$ We would like to modify the bracket polynomial of a knot into a polynomial not to change by Reidemeister move I.

The 3rd period

(i) Let's make a polynomial of a knot which not change by Reidemeister move I. The X-polynomial is obtained! $\Rightarrow$ Let' find the same pair in

following 4 tangles: ⟋ , ⟍ , ⟋ , ⟍ . In fact, ⟋ and ⟍ are different and ⟋ and ⟍ are different. However, if we look only around their crossings of these tangles, the crossings ⤬ and ⤬ are the same

crossings. By this reason, it is necessary for us to put an orientation to classify

these tangles as follows: ⤬ and ⤬ . Then we can see that these

crossings are different. Thus, when we put an orientation on these tangles an

orientation, we can classify these tangles even if we look only around their

crossings of these tangles. ⇒ Let's give an orientation on a knot and find two

types of crossings. Note: We defined the sign of ⤬ to be +1 and the sign

of ⤬ to be -1. Even if we put the reverse orientation on the knot, these signs are

unchanged. Let $w(L)$ be the sum of the signs of all crossing points of L which we

call the *twist number*, which is independent of a choice of the orientations of a

knot L. The definition of the X-polynomial $X(L)$ for a knot L is given by

$$X(L)=(-A^3)^{-w(L)}<L>,$$

The X-polynomial $X(L)$ is unchanged by the Reidemeister move I. ⇒ Let's

calculate the X-polynomials of ⬡ and ⬡. ⇒ We can see

$$X(⬡) = -A^{16}+A^{12}+A^{4}, \quad X(⬡) = -A^{-16}+A^{-12}+A^{-4},$$

and we can see that the mirror image ⬡ of ⬡ is not the same as ⬡.

Photo 6.3.1

6.3.3. Impression of pupils   We had a *questionnaire* to the 32 students in the class B. The question is as follows:

(1)Check following items

	low		middle		high
· Inclination	1 -	2 -	3 -	4	- 5
· Understanding	1 -	2 -	3 -	4	- 5
· Interesting	1 -	2 -	3 -	4	- 5
· Wonder	1 -	2 -	3 -	4	- 5
· Do you have many questions	1 -	2 -	3 -	4	- 5
※Write your questions concretely :	1 -	2 -	3 -	4	- 5
· I want study more about the knot	1 -	2 -	3 -	4	- 5
· I understand meanings to study mathematical knots	1 -	2 -	3 -	4	- 5

The result of the questionnaire was as follows:

·Inclination	1(2), 2(2), 3(4), 4(13), 5(12)
·Understanding	1(3), 2(2), 3(10), 4(9), 5(8)
·Interesting	1(3), 2(2), 3(9), 4(11), 5(7)
·Wonder	1(3), 2(2), 3(8), 4(11), 5(8)
·Do you have many questions	1(4), 2(3), 3(10), 4(6), 5(9)
·I want to study more about the knot	1(2), 2(2), 3(12), 4(12), 5(4)

·I understand meanings to study mathematical knots

$$1(6), \ 2(6), \ 3(9), \ 4(7), \ 5(4)$$

(2)Impressions of our pupils

<positive impressions>

· Before this practice, I thought that it is very easy, but it was very interesting.

· Like a puzzle.

· I understand a little.

· It is interesting to make some knots.

· I am interested in this practice, but I didn't understand a projection of the knot.

· Before this practice, I thought it is very easy. It is very interesting that a formula represents a knot.

· It is interesting to make a formula of a knot.

· It is interesting for me to make some knots.

· It is interesting to make a knot and try to change a knot into another. I didn't understand to calculate the polynomial of a knot, but these polynomials are very mysterious to me.

· It is interesting to untie some knots.

· It is difficult for me to remember some rules, but I feel that these rules are the same as some of mathematical rules which we know.

< negative impressions >

· The twisting number was difficult to me.

· The X-polynomial was difficult to me.

· It is difficult for me to see the sign +1 or -1 at every crossing.

· Teacher spoke very fast in this class of the last time.

· I need more useful mathematics and I didn't understand the X-polynomial.

· I need more time to see the X-polynomial.

· I understood the definitions of A and B, but I didn't understand any more.

· I don't find the relationship of the bracket polynomial between a knot and the mirror image.

· I saw that knot theory is mathematics, but I don't need this mathematics. I think only specialists study knot theory.

· I don't understand the meaning of making the X-polynomial of a knot.

· I don't understand why knot theory is mathematics.

Photo 6.3.2

Photo 6.3.3

(3)Considerations (class B, 32 pupils)

After the classes, the pupils answered our questionnaires. The conclusion was as follows.

· Very good and good	78%
· Completely understand and understand	53%
· Very interesting and interesting	56%
· Want to know more knowledge of the knot	50%
· Understand the aim of studying the knot	34%
· The knot is mysterious	56%
· The knot is not so mysterious	16%

On the other hand

· It is not so good for me	9%
· I cannot understand	16%
· It is not interesting to me	16%
· I don't want to know about knots	13%
· I don't understand the aim of studying the knot	38%

From these conclusions, we found that many pupils are interested in the mirror image of a knot and the X-polynomial of a knot in senior high school. We feel especially that the pupils who are not good at mathematics were interested in studying a knot by a string. To compute the bracket polynomial, we showed some examples of splicing transformations of knots for the pupils' understanding. Then the pupils understood how to compute the bracket polynomial. However, we found that it is something difficult for the pupils to understand how to use the rule ③ and when they use addition or multiplication in a splicing transformation of a knot. We think that we should emphasize the reason of addition or multiplication for the splicing transformation of a knot. Some pupils who can make the bracket polynomial had very good impression on the knot. However, we think that they don't understand a meaning of the bracket polynomial. Since we cut short our program on the 2nd period, we couldn't explain to the pupils a meaning of the bracket polynomial enough. On the other hand, a pupil who can't compute the bracket polynomial of a knot don't have a good impression on the knot. Since we have only a little time to teach a knot, it was inevitable for us. We need more time to teach a knot to pupils.

6.3.4. Conclusion

We tried to teach knot theory to senior high school students. We took up two contents in knot theory. One is the mirror image of a knot and another is the X-polynomial of a knot. They are very interesting for us and we thought that high school students can understand these contents. However, we have only two periods to teach these contents. So we couldn't teach how to make the X-polynomial to the pupils. We only taught the pupils how to compute the bracket polynomial. The X-polynomial is an invariant of a knot, but the bracket polynomial is not. It was inevitable for us. We introduced the mirror image of a knot and let the pupils find whether or not the mirror image of a knot is the same as the original knot. We can see that the trefoil knot and the mirror image are actually not the same if we can compute their X-polynomials.   Many pupils were interested in the mirror image of a knot. In particular, the *figure-eight knot* is a useful teaching material in high school. We want to have the pupils finding that the mirror image of the figure-eight knot is the same as the original knot by using a string. In the 2nd period, we let the pupils compute the bracket polynomial. It will be a motivation to pupils' computing the X-polynomial to find whether or not the mirror image of a knot is the same as the original knot. We want to have the pupils being interested in how a knot is represented by the bracket polynomial. This is a main theme in this practice. To have the pupils compute the bracket polynomial, we used a way to draw a picture of a splicing transformation of a knot. The pupils can obtain the bracket polynomial by drawing this picture. It will be the first time for the pupils to study this kind of mathematics in school. We think that this way is useful to consider a mathematics education from another point of view. In other words, teaching knot theory will be useful to tell pupils mathematical activities through a simple work.   We notice that we must not use many printed sheets in this class. We need a long time to teach it in high school, but we hope it to be taught in many high schools.

References

[1] A. Kawauchi et al, Knot Theory (in Japanese), Springer Verlag, Tokyo, 1990. (English expanded version: A Survey of Knot Theory, Birkhäuser Verlag, 1996.)

[2] C. C. Adams, The Knot Book, W.H. Freeman and Company, 1994. (Japanese version translated by T. Kanenobu, 1998.)

[3] S. C. Carlson, Topology of Surfaces, Knots and Manifolds, John Wiley & Sons, Inc., 2001 (Japanese version translated by T. Kanenobu, 2003.)

[4] A. Kawauchi and T. Yanagimoto et al, An Approach to Teaching Knot Theory in

School Mathematics for Pupils and Students (in Japanese), Project of Teaching Knot Theory in School Mathematics, Research Report as Educational Action in 21st Century COE Program "Constitution of wide-angle mathematical basis focused on Knots(Osaka City University)" Vol.1 2005, Vol.2 2007, Vol.3 2009.

[5] K. Iwase, Mathematical Knots as the Teaching Material in Senior high school, Bulletin of the Tennoji Junior & Senior high school Attached to Osaka Kyoiku University, No.50, 2008.

[6] K. Iwase, Mathematical Knots as Teaching Material in Senior high school - X-polynomials and Mirror images-, Bulletin of the Tennoji Junior & Senior high school Attached to Osaka Kyoiku University , No.52, 2010.

6.4   Education practice in senior high school attached to Osaka Kyoiku University (2)

In this section, we report our practices of teaching some invariants of knots and links through mathematical activities in a senior high school. We consider the possibility and benefits of teaching knot theory through *"mathematical activities"* as follows: After introducing some notions of "invariants" in knot theory such as the linking number, the tri-colorability and the Jones polynomial, students find various properties of knots by experiments and observations, confirm them, generalize their own results and deliver a lecture of their research results. We claim that senior high school students can understand the notion of "invariant" perfectly after learning "Set and Logic". Our research is based on an education practice in Tennoji Senior High School attached to Osaka Kyoiku University.

6.4.1 Introduction and aims   In a junior high school, through studying elementary geometry, students learn mathematical logic. However, they have not studied systematically mathematical logics such as proposition, true statement, false statement, contraposition, converse, conditional statement, and proof by contradiction. Thus, we did not demand them a logical consideration in the practice of teaching on a knot invariant in a junior high school and we had junior high school students consider the tri-colorability of a knot from a brief example, in which we use the trefoil knot and an inductive way of thinking. We briefly review the practice in a junior high school.

Fig.6.4.1   Fig.6.4.2

Junior high school students immediately notice that the left-hand side is not trivial and the right-hand side is trivial in Fig.6-4-1. Next, they color in individual strands by three colors, for examples, white, black, stripe as in Fig.6-4-2. Then, in each crossing, the left-hand side uses three colors and the right-hand side does not so as in Fig.6-4-3. Moreover, they realize that at every crossing, the fact that all three strands are of different colors implies the knot is not trivial and if it is not so, then the knot is trivial. As a result, junior high school students reacted as follows: (1) Teaching materials attracts their interests. (2) They had a fresh surprise that they could judge it by coloring the knot whether a given knot is reduced to a trivial knot or not.

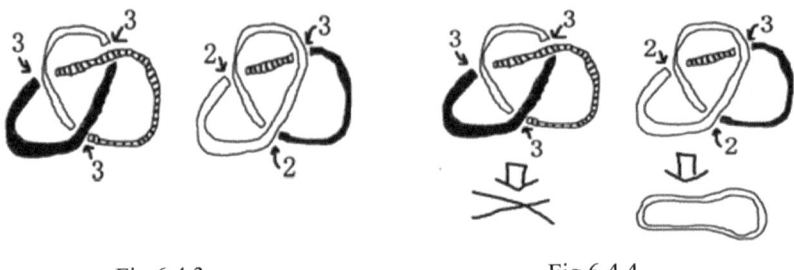

Fig.6.4.3   Fig.6.4.4

Then, in each crossing, the left-hand side uses three colors and the right-hand side does not so as in Fig.6-4-3. Moreover they realize that at every crossing, the fact that all three strands are of different colors implies the knot is not trivial and if it is not so, then the knot is trivial. As a result, junior high school students reacted as follows: (1) Teaching materials attracts their interest. (2) They had a fresh surprise that they could judge it by coloring the knot whether a given knot is reduced to the unknot or not.

In Japan, senior high school students usually learn a field of "Set and Logic" in the second half of the first year in a curriculum of mathematics. We would expect that they had a logical way of thinking into the age group. We thought that we might let them understand *knot invariants* according to the definition. In the previous teaching experiment [Kawauchi, Yanagimoto(2007)], we performed the practice of teaching in the first half of the first year and students of objects did not learn "Set and Logic". A consequence was that they were troubled with logical considerations of knot invariants such as the tri-colorability and the crossing number. Though they took up various problems and discussed them, there were few discussions on the knot invariants. Thus, mathematical experiences from a viewpoint of knot invariants would be better to be taught students after studying "Set and Logic".

The second remark is to examine carefully which knot invariants in knot theory are suitable for teaching materials in which students do mathematics. In New School Teaching Guide for the Japanese Course of Study of Mathematics, the keyword "through mathematical activities" was very seriously taken up. The guideline emphasizes to have students make a conjecture of mathematical properties and rules, and confirm it through further experiments and observations. If they had a conviction that the expectation is true, let they try to prove it. If a counterexample is found, let they make a new expectation through an experiment and observation once again. Thus, students actually do mathematics, so that they obtain good mathematical lessons. We actually consider that some knots invariants in mathematical knot theory are suitable for ones of teaching materials in mathematical activities.

6.4.2   Lesson Plan in senior high school   In the first year of a senior high school, we first taught students a basic knowledge in knot theory.   Then, we let the students decide a theme, and let them study each theme one by one or a group of two or three. They made a resume of each study. In the final lesson, we let the students make a presentation of the theme by using OHP. Throughout this mathematical activity, we shall make clear that knot theory has a variety as the teaching materials.

The target students were 11 boys and 5 girls of the first year in a senior high school.

Teaching content is as follows:

Lesson 1: Let them understand the Reidemeister moves while using strings (2 hours).

Lesson 2: Let them understand notions of a tri-coloring and an n-coloring of a knot as a knot invariant (2 hours).

Lesson 3: Let them understand a notion of Jones polynomial of a knot as a knot invariant (2 hours).

Lesson 4: Decision of a theme and making of a resume (2 hours).

Lesson 5: A meeting for presentations of research results (2 hours).

"Mathematical knot theory" is a new teaching content, and it is not well known what kind of mathematics it is. We explain briefly a teaching content of knot theory through an outline of Lesson 2 done in an actual class.

6.4.3   Teaching content of Lesson 2

(1)   The purpose of Lesson 2.

(i)   To make them understand the notion of a *tri-colorability* to classify knots.

(ii)   To make them understand that the *tri-colorability* is a knot invariant.

(iii) To make them know the notion of a *tri-colorability* on a link.

(2)   The content of Lesson 2:

When we make a Hopf link by using a rope in practice, we easily realize that we cannot deform the Hopf link into a trivial link. However, when we explain it to somebody, we do not understand how we should explain it. In the class of Lesson 1, we teach the students the definition of "linking number" and show that the linking number is invariant under the Reidemeister moves I, II, III. Thereby, we understand that we cannot deform the Hopf link into a trivial link by a finite number of times of the Reidemeister moves. Also, making actually a trefoil knot by using a rope, we sensuously understand that we cannot deform the trefoil knot into a trivial knot. It is difficult to explain why the trefoil knot cannot be deformed into a trivial knot. In the practice of a senior high school, we teach the definition of a *tri-colorability* instead of making the students find the notion of a tri-colorability.

A coloring of a knot is to make colorings of the strands between the under-crossings in a diagram of the knot.

Definition 1. A knot is *tri-colorable* if :

Rule 1: At every crossing of the diagram, either all the three strands are of different colors, or of the same color.

Rule 2: Two or three colors are used in the whole.

Exercise 1. Show that two knots below are tri-colorable.

   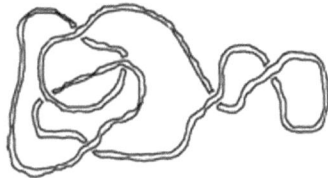

Fig.6.4.5   Fig.6.4.6

(Solution)   Two knots above are both the left-handed trefoil knot.   We color them by white, black and stripe. In Fig. 6.4.7 three colors are used at the three crossings. In Fig. 6.4.8, we find the same color in a twisted part, but we understand that Fig. 6.4.8 is also tri-colorable.

Fig.6.4.7   Fig.6.4.8   □

Exercise 2. Find a tri-colorable knot in the following two knots:

   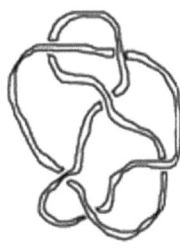

Fig.6.4.9   Fig.6.4.10

(Solution) Fig. 6.4.9 is tri-colorable as follows:
Fig.6.4.10 is not tri-colorable, but considerable trials
and errors are needed for an explanation of the proof.
When students repeat them many times, it comes to be
able to judge early soon whether or not a given knot is
tri-colorable.   □

Fig.6.4.11

Exercise 3.   Show that the knot in Fig. 6.4.12 is
tri-colorable.

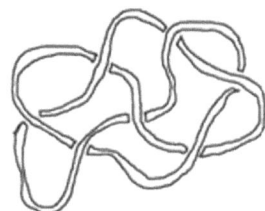

Fig.6.4.12

(Solution) To color this knot, it is necessary for
students to look from different viewpoints. If
students notice that the crossing in a central part
is taken by the same color, then they can do
as in Fig.6.4.13.   □

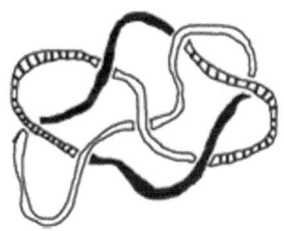

Fig.6.4.13

We are going to show that the Reidemeister moves I and II can be made without
affecting the tri-colorability. As there are many cases to show it for the Reidemeister
move III, we show only one example. The proof of the remaining cases of the
Reidemeister move III is left to the students as a research theme. From this, we derive
the following theorems:

Theorem 1.   The tri-colorability of a knot is a knot invariant.

Theorem 2.   Let K and L be two equivalent knots. If K is tri-colorable, then L is so.

Theorem 3.   Let K and L be two knots. If K is tri-colorable and L is not, then K and L
are not equivalent.

To derive Theorems 2 and 3 from Theorem 1, a knowledge of logic such as a
contradiction or a contraposition is necessary for students. It would be interesting to the
students to confirm that students learned set and logic, because they know that school
mathematics are useful in this case. We think that students understood these theorems to

some extent.

Exercise 4. Is the figure-eight knot *tri-colorable*? Does the left-handed trefoil knot differ from the right-handed trefoil knot by using the tri-colorability?

Exercise 5. By observing the results of Exercise 4, what opinion on knots do you have?

(Solution) The trivial knot is colored by only one color, and hence the trivial knot is not tri-colorable. Therefore, we cannot deform the trefoil knot into the trivial knot absolutely. Also, the figure-eight knot is not tri-colorable, too. This implies that the trefoil knot cannot be deformed into the figure-eight knot. But by using the tri-colorablity only, we cannot know whether the figure-eight knot is deformed into the trivial knot or not. Thus, we need further developments on new invariant of knots.
□

(3) Considerations in Lesson 2.

(i) We let students bring their highlighter and colored pencil to the class. Students enjoyed their work by hand and by trial and error. They spent much time on their work even if the knot is tri-colorable and they can color it by using three colors well. It seemed to be difficult in the case with a crossing of the same color as in Fig.6.4.12. In the case of non-tri-colorable knots, most students could conclude through several cases that the knot is not tri-colorable.

(ii) In the proof of Theorem 1, we showed that it is invariant under the Reidemeister moves I and II only. Though we stated its outline only, there were some students which were interested in the proof of Theorem 1 and they address the problem by using several approach. We mention their research results later.

(iii) In case they studied "Set and Logic", they could accept the proof and the meaning of Theorems 2 and 3. In fact, it follows that Theorem 1 implies Theorem 2. Taking the contraposition of Theorem 2, we have Theorem 3.

6.4.4 Examples on student research results We performed classes from lesson 1 to lesson 3 as in 6.3.3 and we have students determine a theme of knot theory which they feel a great interest in . In lesson 4, we talk over their theme with them to deepen their understanding of knot theory. In lesson 5, we have students announce their research results. For example, their talk titles are as follows:

(i)   On properties of the *tri-colorability*.

(ii)   On properties of the bracket polynomial.

(iii)   Evenness of the degrees of Jones polynomials in all knots and links.

(iv)   The relation between knots and their mirror images in Jones polynomials.

(v)   Finding a new method for a knot invariant.

We state the content of (i) "On properties of tri-colorability" in detail, where (S) expresses a student and (T) a teacher.

Firstly, (S) completes the proof that the tri-colorability is a knot-invariant. In fact, (T) shows only the proofs of the Reidemeister moves I, II and one case of III in the class. (T) leaves the remainder of the proof as a homework. As the proof of III must examine various cases, even the technical book leaves as an exercise to the reader. Since it is difficult to explain it definitely, (S) accomplishes it.

Comment: (T) make students experiment whether many knots are tri-colorable or not. Particularly, (S) had a trouble with a case with a crossing of one color. At last, (S) could paint by three colors in a response to advices from some other students. (S) asks the following problem from such an experience.

"Is it possible to tri-color a knot along one way?" (S) proposes the following method: Decide a base point in the knot properly and advance the knot in one direction, and one change a color in turn whenever it meets a crossing such as red, blue, yellow, red, blue, ... We decide to perform this operation automatically.

Comment: The critical mind is very interesting. As we think that there does not exit how to paint so regularly, we do not do such an expectation. As (S) had a trouble with painting with three colors, (S) searches how to color more simply. Of course, it is not easy. (S) have understood it. However, (S) goes on with his work. One of the important points is as follows: (S) asks: Is "the tri-coloring in this way" invariant?

(Answer): For a twist, a color must not change, but a color changes forcibly in this definition. Therefore, this coloring is not invariant in the Reidemeister move I.

In the Reidemeister move II, the left-hand side figure is tri-colorable, but the right-hand side figure is not tri-colorable by this coloring.

Therefore, regrettably, (S) says that this coloring method is not a knot invariant and it is difficult to find a new invariant.

Comment: After (S) explains definitely that the tri-colorability is a knot invariant, (S) says the next dissatisfaction: Actually when it is colored by three colors, trial and error are considerably needed. There is a case that one cannot color by three colors when one begins to color properly. Thereby (S) searches how to color more simply. We felt that the idea of "tri-coloring in one way" was very interesting, though unfortunately it is not a knot invariant. (S) appeared to be very disappointed. (T) understood this situation when (T) called him to know a progress on (S)'s study. In this situation, his announcement is easily over. (T) asked (S) whether there is a knot to be able to color by this method well or not. (S) reported (T) that there exists an example of knots on a ring to be able to color well. Then (T) asked further (S) as follows. When does the knot have a property of tri-colorability? (S) expected that the knot is tri-colorable only when the number of crossings is a multiple of 3. (S) investigated several cases and proved that it is a necessary and sufficient condition on such knots. We try to follow an argument of (S) as follows:

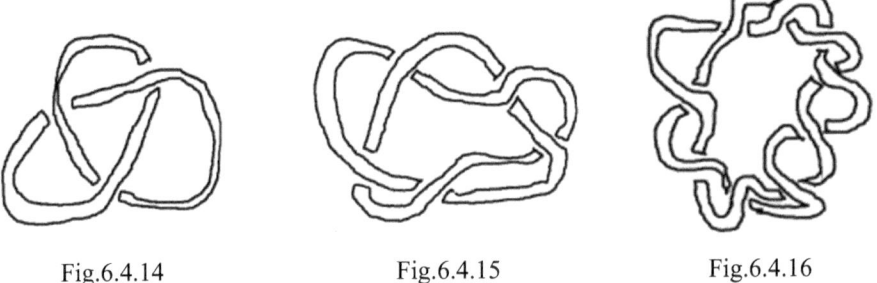

| Fig.6.4.14 | Fig.6.4.15 | Fig.6.4.16 |

Note that the knot in Fig.6.4.14 has three crossing points, the knot in Fig.6.4.15 five and the knot in Fig.6.4.16 nine. Then they are not a trivial knot. These knots seem to put the following forms together.

Fig.6.4.17

First, we examine a property on these forms. We paint them with three colors.

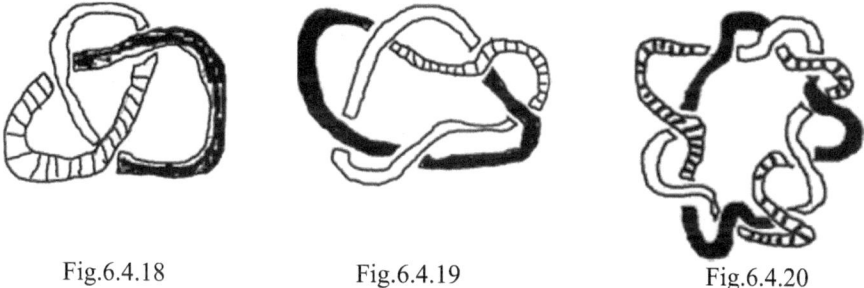

Fig.6.4.18   Fig.6.4.19   Fig.6.4.20

The knots in Figs.6.4.18 and 6.4.20 are tri-colorable but the knot in Fig.6.4.19 is not. We examine whether the following knot with seven crossing points is tri-colorable as we expect that a knot with even numbers of crossings has the same property.

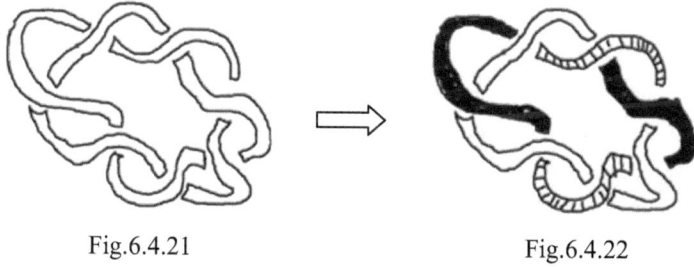

Fig.6.4.21   Fig.6.4.22

The knot is Fig.6.4.21 is not tri-colorable as shown above and similarly the knot with eleven crossing points is not. Next, we examine the case of links.

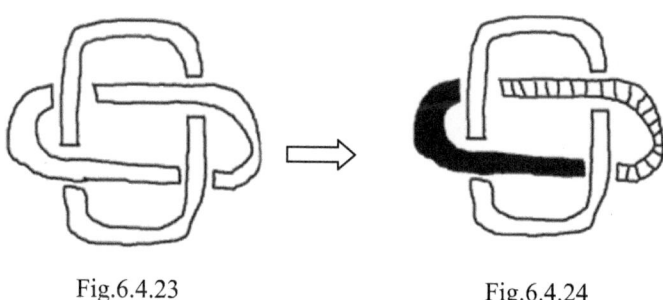

Fig.6.4.23   Fig.6.4.24

The link with four crossing points in Fig.6.4.23 is not tri-colorable. The link with six crossing point in Fig.6.4.25 is tri-colorable as shown below and the links with eight and ten crossing points are not tri-colorable:

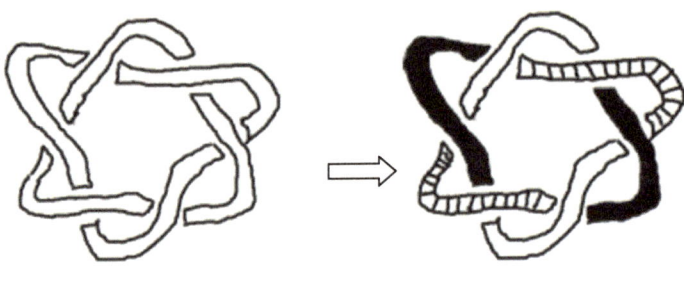

Fig.6.4.25   Fig.6.4.26

Then (S) noticed the following conjecture by many experiments on knots and links with many crossing points:

Theorem.   A knot or link on a ring (=a closed 2-strand braid) is tri-colorable if and only if the number of the crossings is a multiple of 3.

Proof.   A strand of the knot diagram is colored by blue. Then the color of the second strand is blue or another color. If it is blue, then the next color becomes blue automatically, and all the colors become blue. This is not the case. Therefore, the color of second strand is red. The next becomes green necessarily. As a result, it is colored in turn such as "blue⇒red⇒green⇒blue⇒red⇒green⇒..." To show that the knot is tri-colorable, we pay our attention to the last strand. As the first strand is blue, the color leading to the color of first strand must be green. Because of the order of blue, red and green, we can regard these 3 as one group. Then we know that it is green in a multiple of 3. Hence it has a multiple of 3 strands if and only if it has a multiple of 3 crossings. QED

Photo 6.4.1

Photo 6.4.2

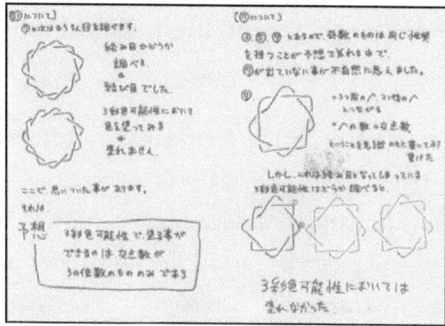

Fig.6.4.27

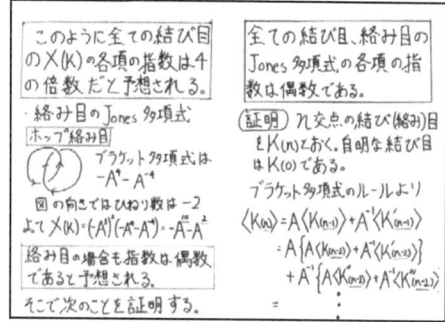

Fig.6.4.28

6.4.5 Conclusion In this section, we have attempted to find the possibility and benefits of teaching knot theory in a senior high school. We can conclude that "knot theory" is flexible teaching materials in which students can have good experiences of mathematical activities such as (i) Experiment-Observation⇒(ii) Conjecture ⇒(iii) Confirmation-Proof ⇒ (vi) Generalization. Even if students are not good at mathematics and cannot understand the subject with knot theory on his/her own, we can offer students a subject from knot theory in each level of ability and moreover make them purse it deeply. Though knot invariants in knot theory made a difficult impression on students, they had a little confidence by the fact that they can obtain a result by themselves. This fact represents the variety of knot theory. Whereas we gave students only three teaching materials of knot invariants such as the linking number, the tri-colorability and the Jones polynomial, students made various research results. This shows that knot theory has rich contents in mathematics education. Students delivered a lecture on mathematical results to others by using OHP, and it brought them a valuable experience, because there are few opportunities in school mathematics such that students find a subject by themselves, solve it and deliver a lecture on their results to others. As an impression of teaching knot theory, we realized that senior high school students can understand a notion of knot invariants perfectly.

References

[1] A. Kawauchi et al, Knot Theory (in Japanese), Springer Verlag, Tokyo, 1990. (English expanded version: A Survey of Knot Theory, Birkhäuser Verlag, 1996.)

[2] T. Yanagimoto, Y. Seo, K. Iwase, M. Terada and R. Kaneda, An Approach To Teaching Knot Theory in Schools, Proceeding Fourth East Asia Regional Conference on Mathematics Education (EARCOM4), 411-416, 2007.

144

6.5 Cooperative education practice between senior high school and university

This report is based on a report written when working for Osaka prefectural Tennoji high school. Osaka prefectural Tennoji high school was selected as a *Super Science High School* (SSH) of the Ministry of Education in 2004. A coordinated education with the university was planned as one of the researches in SSH in 2005, and it cooperated with Osaka City University.

6.5.1 Cooperation of Tennoji high school and Osaka City University The third graders of the science and mathematics department of the Tennoji high school became able to attend "Introductory lecture on mathematics" opened in 2006 for the first grade students of Department of Mathematics of Osaka City University. The theme in this year was "Mathematics of knot" and A. Kawauchi and T. Kanenobu, who are specialists of knot theory and professors at the graduate school of science, took turns giving this lecture. The high school students and I attended this lecture once a week at Osaka City University. This lecture was an introduction to "Knot theory" that needed a high school level preliminary knowledge of mathematics. It was a 90-minute lecture that applied about 30 minutes to the review and the practice and gave very careful and attentive training. This made the students accustomed to diagrams of "Knot" and "Link" and trained them well enough to be familiar with the diagrams. It also helped the students to develop their mathematical intuition, to understand Reidemeister moves and the meanings of knots and links and to be able to draw their own new diagrams of knots and links through a number of exercises. The students also became able to make calculations on the twisting number of a knot diagram or a link diagram and the linking number of a link. In addition, they learned the bracket polynomial of a knot diagram or a link diagram.

6.5.2 Students during the lectures At first, the students were puzzled that the lectures were quite different from the classes of the high school, and felt nervous. However, since "knots" are found everywhere in our daily life, and the lectures were given in a plain way, they gradually became accustomed to the lecture of the university and understood the contents of the lecture. At the end of each of the lectures, I often saw some students questioning the lecturer on the contents of the lectures. The following photograph shows the students during the lectures. The university students also sit at their desks, but are not in this photograph.

Photo 6.5.1: Students during class

The following photograph shows the students asking Prof. Kawauchi some questions after the lecture.

Photo 6.5.2: Students asking Prof. Kawauchi

The following two photographs show scenes of lectures by Kawauchi and Kanenobu.

Photo.6.5.3: A lecture by Kawauchi   Photo.6.5.4: A lecture by Kanenobu

146

The following photograph shows a student trying an exercise.

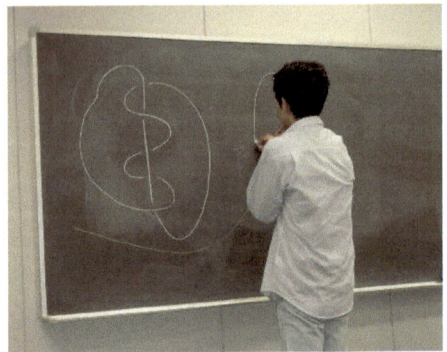

Photo.6.5.5: A student trying an exercise

6.5.3 A research into students' mathematical ideas   When there was no lecture at the university, I had classes at the high school.   The following knots in Fig.6.5.1-6.5.9 show some of the works that the students made in order to understand the Reidemeister moves during the lesson.   They showed that the knot in Fig.6.5.1 can be transformed into the knot in Fig. 6.5.9 by the Reidemeister moves, so that the knot in Fig. 6.5.1, which seemingly looks complex, is actually a trivial knot.

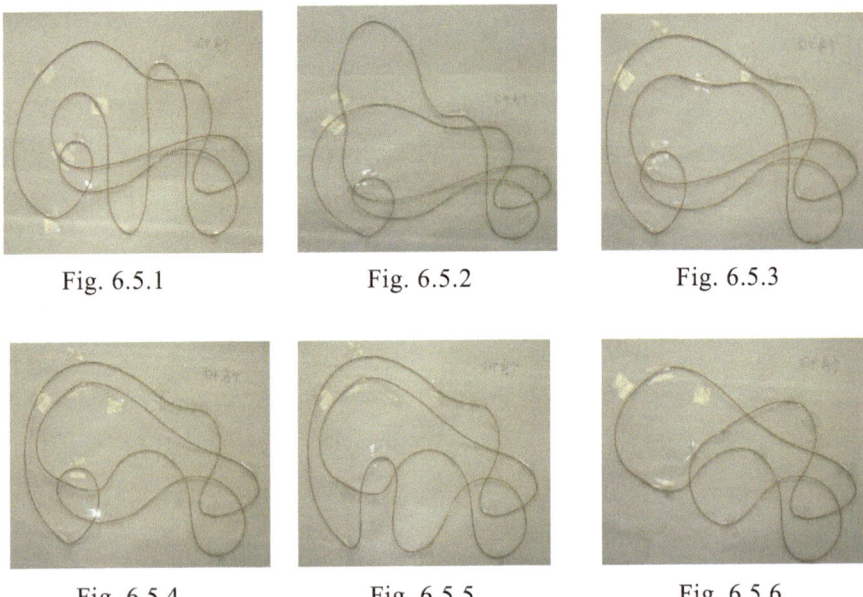

Fig. 6.5.1   Fig. 6.5.2   Fig. 6.5.3

Fig. 6.5.4   Fig. 6.5.5   Fig. 6.5.6

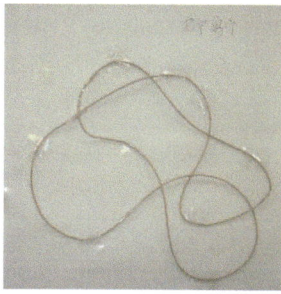
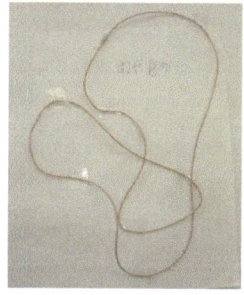
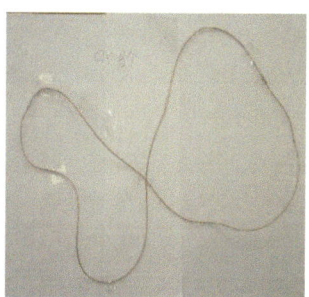

| Fig. 6.5.7 | Fig. 6.5.8 | Fig. 6.5.9 |

The students became able to understand the concept of a knot or link, etc. on their own powers by making a concrete thing like those shown in the photographs (Fig.6.5.1- 6.5.9).   Photo.6.5.6 shows a classification table distributed in the lecture at Osaka City University.   The following photographs (Photo.6.5.7 -6.5.9) show the students in my class.   They are making one of the knots in a knot table by using a paper tape.

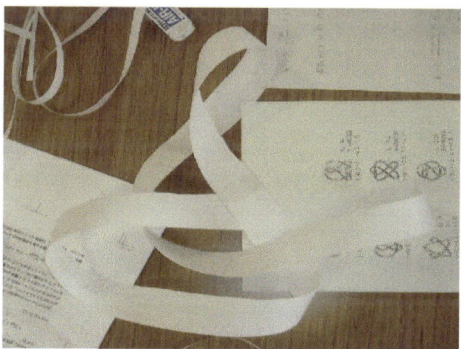

Photo.6.5.6: A knot table and work

| Photo.6.5.7 | Photo.6.5.8 | Photo.6.5.9 |

In one of my classes, I made the student make a *Moebius strip* and cut it along the belt. Then the students made tried various ways to cut it along the belt. Among the works the students made in the class, there were some works resembling some knots in the knot table. Later, in the university lecture, they questioned Kawauchi on it.

In the meantime, one of the students, a captain of our gymnastic club, found that a fabric tape used in performing on the horizontal bar was shaped like a Moebius strip. When he tried to repair the tape, he wondered which way he should give a twist to it, rightward or leftward. He thought that this question might be the same kind as those in my class, and told the other students on the fabric tape. In this way, the students started to think of the question asking whether "Moebius strip of a right twist" and "Moebius strip of a left twist" were the same or different.

They thought they could apply knot theory to solve the question of those "Moebious strips." They tried various ways to cut the paper belt and asked a lot of questions to Kawauchi, and finally they learned that one of the ways to cut the belt (cutting it on both side along the belt) could be related to "links" in knot theory, ant that it was effective to apply the idea of direction. As a result, they completed the proof of a topological difference of a right twist and a left twist.

The research of this "Moebious strip" won the 1st place in the mathematics section in "the Symposium of a fiscal year 2006" of the Ministry of Education, Culture, Sports, Science and Technology held on August 9th and 10th, 2006 in Pacifico Yokohama, which was sponsored by the independent administrative agency, Science and Technology Promotion Organization. In addition, this research won the highest award in the four "first prizes" of the sections of mathematics, physics, chemistry and biology.

Their research summaries are shown as follows:

"Consideration of Moebius strip"

1. Purpose of the research

We think of a question "Is there a topological difference between the left-handed twist Moebius strip and the right-handed twist Moebius strip?"

2. Research Method

Consider a link which consists of the boundary loop "k" with any orientation of a Moebius strip "M" and an oriented loop "k'" which is parallel to "k" in "M". Then the linking number Link (k, k') distinguishes the difference between the right-handed twist Moebius strip and the left-handed twist Moebius strip.

3. Result of the research

The left-handed Moebius twist strip and the right-handed twist Moebius strip are shown in Fig. 6.5.10 and Fig. 6.5.11, respectively.

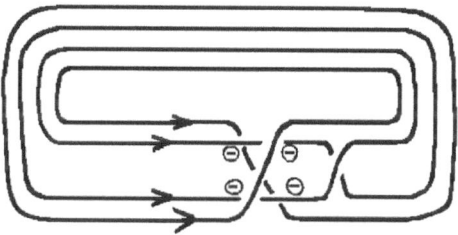

Fig. 6.5.10 Left-handed twist

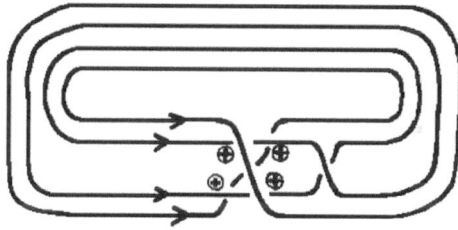

Fig. 6.5.11: Right-handed twist

For the left-handed Moebius strip, the sum of intersection signs is -4 and hence the linking number Link $(k, k') = -2$.   For the right-handed Moebius strip, the sum of intersection signs is +4 and hence the linking number Link $(k, k') = +2$.   Since Link (k, k') is a topological invariant and is invariant under changing both the orientations of "k" and "k'", we can conclude that the left-handed Moebius strip is topologically different from the right-handed Moebius strip.

7. Education Practice at the University as Liberal Arts and Teacher Education

In this chapter, an education practice at the university as a liberal art and a teacher education is introduced. Target students in the education practice were students who become elementary school teachers. A textbook content on basic ideas of knot theory for an education practice is also introduced, which is based on some education practices at universities and colleges as a liberal art and a teacher education.

7.1 Educational meaning

The educational meaning of teaching knot theory in school has the following five points:

1. Fostering students' abilities of a spatial cognition by handlings of mathematical knots.
2. Giving an opportunity for students to learn a concept of invariants.
3. Giving an opportunity for students to discuss a mathematical idea by using some simple concrete examples on knots.
4. Giving an opportunity for students to create a mathematical idea without much preliminary knowledge.
5. Exhibiting significances of learning mathematics. Because knot theory is applied not only to mathematics but also to different sciences such as biology, the high polymer chemistry, physics, psychology, and so on. It is a theory of the forefront being watched in the world.

7.2 Knot Theory in School Mathematics as liberal art and teacher education

The technical book of knot theory is for a technical researcher training. To make a textbook for a liberal art and a teacher training, the following elements will have to be counted at least.

1. A preliminary knowledge is not needed for the student.
2. The student realizes happiness to build mathematics.
3. The ability to teach it as a child education is cultivated.

4. Intrinsic notions of knot theory such as the equivalence of knots and the classification of knots are taken in school mathematics.

7.3 Syllabus planning

The education practices were carried out in Kobe Shinwa Women's University and Shukugawa Gakuin College from 2006 to 2009.

A central argument in the education practice in 2006 is which knot can be untied. As for the tri-colorability, if it is untied, then it does not have the tri-colorability. However, even if it was not untied, it happens not to have the tri-colorability. Thus, a confusion arose in the arguments on the tri-colorability and the untying. In the education practice in 2007, though untying a knot is also well discussed at the introduction stage, a central argument was shifted to the equivalence of knots. They learned the composition of knots newly in the third hour. A link and the linking number were learned in the fourth hour, and the tri-colorability was learned in the fifth hour. In the education practice in 2008, much time was spent on learning of the bracket polynomial but Jones polynomial. An examination is further needed because there are teaching gaps between the linking number, the tri-colorability and the bracket polynomial. The content of the education practice in 2009 is the same as the content of the preceding fiscal year, but it ended in learning of the linking number and the tri-colorability to avoid a confusion. The following syllabus planning was made with considerations stated above.

1. Untying a knot. (1 hour)
2. Reidemeister moves (1 hour)
3. Composition of knots (1 hour)
4. Link and linking number (3 hours)
5. Tri-colorability (2 hours)
6. The bracket polynomial (4 hours)
7. The Jones polynomial (2 hours)

7.4 The outline of the practice

10 educational practices as liberal arts or teacher education were done as follows:

Education practice ① was carried out as "Special Seminar on Childhood Education I" at Kobe Shinwa Women's University in 2006. The target students were four 3rd graders in a teacher training course to be elementary school teachers.

Education practices ② and ③ were carried out as a subject of liberal arts "Mathematics re-discovery" at Shukugawa Gakuin College in the first and the last semesters from April, 2006 to March, 2007. The target students were seven 1st graders

Based on the above practices, the following education practices were carried out. Education practice ④ at Kobe Shinwa Women's University from May to October, 2007: A 3rd grader in a teacher training course learned in a subject "Specialized Seminar on Childhood Education I "

Education practice ⑤ at Shukugawa Gakuin College from October, 2007 to January, 2008: Three 1st graders in a teacher training course learned in a subject "Mathematics re-discovery"

Based on the above practices, the following education practices were carried out. Education practice ⑥ at Kobe Shinwa Women's University in 2008: Two 1st graders and one of my seminar students in the 4th grade learned in a subject "Mathematics".

Education practice ⑦ at Shukugawa Gakuin College in the first semester of 2008: Four 1st graders and a 2nd grader learned in a subject "Mathematics re-discovery".

Education practice ⑧ at Shukugawa Gakuin College in the last semester of 2008: Eight 1st graders and five 2nd graders learned in a subject "Mathematics re-discovery".

Education practice ⑨ at Shukugawa Gakuin College in the first semester of 2009: Eleven 1st graders learned in a subject "Mathematics re-discovery".

Education practice ⑩ at Shukugawa Gakuin College in the last semester of 2009: Six 1st graders learned in a subject "Mathematics re-discovery".

7.5 Considerations

In the following item, considerations based on the practices as a liberal art education and a teacher education at universities and colleges are explained.

7.5.1 Students had interests in untying knot and link.   Students came to think of the conditions of untying gradually by untying many knots and links. Then, they had interests to create mathematics.

7.5.2 Reidemeister move II is firstly noticed, and secondly Reidemeister move I is noticed. However, Reidemeister move III where the crossing number is unchanged is not noticed. The first untying condition of a knot diagram with 2 crossing points is that there is an upper path containing 2 crossing points in the diagram, where the Reidemeister move II is used to untie it. The second untying condition of a knot diagram with 2 crossing points is that any upper path does not contain 2 crossing points, where the Reidemeister move I is used to untie it. It may be noticed in an early stage of untying to reduce the number of crossing points, because students only move crossing points in the process to untie the knot and do not notice that the number of crossing points is unchanged.

7.5.3 It is meaningful to learn invariants in the order from the linking number to the tri-colorability. In 2006, students learned invariants in the order from the tri-colorability to the linking number in accordance with the educational practices of the elementary school and the junior and senior high schools. After 2007, the order was changed from the linking number to the tri-colorability, because this order is standard in the technical book of knot theory. The education practice in 2006 was based on a knot (that is, a link with one component) which appears simpler than a link of 2 components. Then it was understood that coloring a knot is simpler than calculating the linking number. Although they computed the linking number of a link and painted a knot, it was thought by the practices at the university after 2007 that calculating the linking number is not always difficult for the university students. After the untying condition of a knot and the Reidemeister moves were examined, students learned the tri-colorability and had interests in relations with coloring and untying of a knot. A tri-colorable knot cannot be untied, but a non-tri-colorable knot need not be untied, where a confusion made occur. By this reason, the tri-colorability was used for the equivalence of knots rather than the untying. The equivalence of knots is a main object and the invariant is a main target of the consideration in the research of knot theory. Education practices of knot theory at universities and colleges as a liberal art and a teacher education were thought to adopt the same object and target, too. The linking number of a trivial link is 0. Then the linking number of Whitehead link is 0. The absolute value of the linking number of Hopf link is 1 and it can be judged that the Hopf link is different from the trivial link. On the other hand, the difference between the trivial link and the Whitehead link cannot be decided by the linking number. By this reason, students had interests when the teacher suggested that they had to think of another invariant. It can be judged that it is different from trivial link and Whitehead link by examining the tri-colorability. This is a meaning that the tri-colorability is developed after the linking number, so that students recognized it here, too.

7.5.4 On link and linking number   Students do not have any idea to attach an orientation to a link. It an examination subject whether or not an orientation should be easily attached to a link. Students say that a link is easier understood if a link is defined to be an oriented link since students only compute the linking number according to definition (though they mistake numerical values and get confused). Knot and link were untied, and Reidemeister moves were examined, and it had been discussed smoothly here. There is a difficulty in taking the value ± 1 in the calculations of the linking number. Even if it is calculated carefully, the value ± 1 is mistaken. The linking number is a bottleneck.

7.5.5 On tri-colorability   Students examined a condition of the tri-colorability and understood well a meaning of the tri-colorability. It needs a discussion whether or not a given knot is seen to be untied from the tri-colorability. What can be said is only that a trivial knot is not tri-colorable. This gets confused if logic is not put in order.   As the contraposition " A tri-colorable knot cannot be untied."   The tri-colorability can be also applied to any equivalent knot together with the same tri-colorablity. As the contraposition, "Two knots are not the same knots if their tricolorilities do not coincide." It can be said that an investigation on the equivalence of two knots is better than an investigation of untying a knot, as an application of the tri-colorability.

7.5.6 On bracket polynomial   Students were interested in that the crossing points of a knot diagram was changed into a diagram without crossing points by splicing. Students were surprised and interested in introducing a polynomial to a knot. However, the hurdle of the calculation was very high because some students cannot compute polynomials. Thus, the teacher had to show the students a meaning of why the bracket polynomial is needed. As for this reason, the decision that the trivial knot is different from the figure-eight knot cannot be made by the tri-colorability. Hence we must think of another invariant. Students had interests when the teacher explained that it could be shown by computing the bracket polynomials of the trivial knot and the figure-eight knot. Furthermore, students had interests in what happens for the bracket polynomials of a knot and the mirror image. If a normalized bracket polynomial is not introduced, then the calculation of the polynomial finishes in a halfway. Then the polynomial is not sometimes 1 even if the knot is a trivial knot. The normalized bracket polynomial is the product of $\left(-A^3\right)^{-w(L)}$ and the bracket polynomial, where $w(L)$ denotes the writhe, namely the sum of crossing signs, of an oriented knot or link diagram L. It was explained as follows why such a product was made: As for the *Reidemeister moves* II and III, the bracket polynomial is unchanged. However, as for the Reidemeister move I, it is changed and we must seek a condition to be unchanged in the Reidemeister move I.

$$L'' \qquad\qquad L \qquad\qquad L'$$

For the knot diagrams L, L' and L'' in the figures above whose hidden parts denote the identical diagram except the figures above, we obtain the writhe $w(L)=\omega$, $w(L')=\omega+1$ and $w(L'')=\omega-1$ for an integer ω. Further, we have

$$\langle L'\rangle=\left(-A^3\right)\langle L\rangle, \quad \langle L''\rangle=\left(-A^{-3}\right)\langle L\rangle=\left(-A^3\right)^{-1}\langle L\rangle. \quad \text{(Refer to p.176)}$$

When $-A^3=\alpha$ is taken, they are $\langle L'\rangle=\alpha\langle L\rangle=\alpha^1\langle L\rangle$ and $\langle L''\rangle=\alpha^{-1}\langle L\rangle$ here.

By the way, they are $\alpha^{-(\omega+1)}\langle L'\rangle=\alpha^{-(\omega+1)}\cdot\alpha^1\langle L\rangle=\alpha^{-\omega}\langle L\rangle$ and

$$\alpha^{-(\omega-1)}\langle L''\rangle=\alpha^{-(\omega-1)}\cdot\alpha^{-1}\langle L\rangle=\alpha^{-\omega}\langle L\rangle.$$

Therefore, it is understood that the normalized bracket polynomial is unchanged in the Reidemeister move I as well as it is unchanged in the Reidemeister moves II, III. In this argument, putting $-A^3=\alpha$ would be easier to understand. Thus, the normalized bracket polynomial $N_L(A)$ of a knot or link L is defined as follows.

$$N_L(A)=\left(-A^3\right)^{-w(L)}\langle L\rangle$$

The hurdle of this argument is very high. A power to think of mathematics is seriously requested here. It may be thought that the trivial knot and the figure-eight knot are judged to be different at the stage that the bracket polynomial is confirmed to be unchanged under the Reidemeister moves II, III without introducing the normalized bracket polynomial, for the bracket polynomial of the trivial knot and the figure-eight knot are 1 and $\dfrac{1}{A^8} - \dfrac{1}{A^4} + 1 - A^4 + A^8$, respectively.

7.5.7 On Jones polynomial   The normalized bracket polynomial in the variable A is nothing but Jones polynomial in the variable t by taking $A = t^{-\frac{1}{4}}$. We consider here it as a polynomial in x by taking $A^2 = x$ by the proposal of Kawauchi to avoid a radical sign calculation. A knot diagram is changed into a diagram without crossings called a state, by splicing from which the bracket polynomial is derived. Since there are 3 crossing points in a trefoil knot diagram, we obtain the 8 states from the trefoil knot diagram. Since there are 4 crossing points in a figure-eight knot diagram, we obtain the 16 states from the figure-eight knot diagram. Generally, we obtain the 2^n states from a diagram with n crossing points. If the number of the crossing points of knot diagrams increases, then the number of the state increases exponentially, and the calculation of the bracket polynomial becomes complicated. For a computation of the Jones polynomial, a skein relation on the Jones polynomials is introduced, and the Jones polynomials of knots and links are calculable recursively. Because a hurdle of this computation was considerably high, this learning was practiced only for the students of in a seminar who have mathematical knowledges and are interested in mathematics.

7.5.8 On DNA   DNA has a structure of double helix. A circular DNA is considered as a knot. A rearrangement of DNA can be done with a kind of enzymatic function. Although the teacher explained that a rearrangement of DNA could be done like a crossing change in a knot diagram, it is a future examination subject to adopt a relation between DNA and knot theory as a subject.

7.5.9 On an intention of the education practice at the university   There are two intentions in the education practices of knot theory in school mathematics at universities and colleges. The first intention is the following point.

What kind of an education content can we build from school mathematics of knot theory as mathematics of a liberal art for a general student?   Such education contents are thought to have been almost built by the education practices done until now which are briefly explained as follows. A condition for a knot to be untied is first considered. Then the Reidemeister moves are introduced. The linking number is introduced as an invariant. The linking numbers of a trivial link and a Whitehead link are 0 together and it cannot be judged that they are different. Thus, the tri-colorability is introduced. The tri-coloring of a trivial link is possible as the result. However, the tri-coloring of the Whitehead link is impossible. Therefore, it can be judged that these are different. On the other hand, the tri-colorings of a trivial knot and the figure-eight knot are both impossible. Therefore, it cannot be judged that they are different. We must think of a new invariant here. Thus, the bracket polynomial is introduced. The 16 states of a figure-eight knot are constructed and the bracket polynomial of the figure-eight knot is computed. It can be judged that the figure-eight knot is different from the trivial knot as the result. Even general students can learn the contents until here. Furthermore, students with some mathematical knowledge can learn the normalized bracket polynomial and the skein relation of the Jones polynomials. The Jones polynomials of various knots and links and then complicated knots and links can be also calculated recursively. The second intention is the following point.

Is it possible for the student who learned knot theory in school mathematics to teach knot theory for the children of the elementary school or the junior or senior high school ? Should the student learn what kinds of contents in knot theory to do it? In the case ① of the education practices, one of the students, Kawashima tried an education experiment on knot theory to one schoolchild as a graduation research and collected the research results in her graduation thesis. There were preceding researches in the elementary school and the junior and senior high schools, but the education practice of ① is done for the first time as the university education practice, where the education contents were not yet definitely determined. The student of the education practice ① was the teacher's seminar student and she was one of the students with a mathematical knowledge. The education practice ① progressed with the following contents. The condition for a knot to become untied, the Reidemeister moves, the tri-colorability, the linking number and the bracket polynomial are considered. The trefoil knot were dealt with the bracket polynomial. Next, the normalized bracket polynomial was learned and then skein relation of Jones polynomial was learned. The Jones polynomials of the trefoil knot and the figure-eight knot were calculated. Furthermore, the Jones polynomial of the mirror image of a knot was found. The Jones polynomial was found by using the radical sign by the first education practice. As mentioned above, Kawauchi proposed a way of calculations without using the radical sign, and it was adopted in the education practices after 2007. The students learned the

Jones polynomial from the condition for a knot to be untied in the education practice ①. Furthermore, the teacher introduced the preceding researches of knot theory in school mathematics in the elementary school and the junior and senior high schools. This gave a motivation to the student (Kawashima) who tried an education experiment to the schoolchild. Kawashima made a teaching plan based on "Teaching knot theory in elementary school" (see Vol. 1 of [2]). The textbook content was made by her own work. She tried an education experiment to one schoolchild. Furthermore, the learning teaching plan is made for six hours, whose theme was "Untying a knot". When she becomes a teacher of the elementary school, she can use her graduation research results, and the teacher would like to expect her trying to teach knot theory in the elementary school.

7.6 The contents of the textbook

First of all, there are many interesting stories on history and applications to sciences in knot theory which will attract students. Although their contents are omitted here, several topics borrowed from reference books, etc. should be explained to the students.

A knot (Fig.7.1) is made with one string, whose closing knot (Fig.7.2) cannot be untied. Namely, it cannot be deformed into a circle. A *knot* is a simple closed curve in the 3-dimensional space and a *trivial knot* is a knot which can be untied. It is a target of the argument which knot can be untied.

Fig.7.1   Fig.7.2

The knot in Fig.7.3 is a trivial knot. The knot in Fig.7.4 is also a trivial knot since it is obtained from the knot of Fig.7.3 by twisting.

Fig.7.3   Fig.7.4

A *knot diagram* is the image of a knot by the projection from the 3-dimensional space

160

onto the plane which has only transversely intersecting double points together with upper-lower crossing information. For example, the knot diagram in Fig.7.4 has just one *crossing point.*

7.6.1 Untying a knot   First, examine Problems 7.1-7.5.

Problem7.1. Can the knots ① and ② be untied?

Fig.7.5   Fig.7.6

Problem7.2. Can the knot ② be deformed into the knot ①?

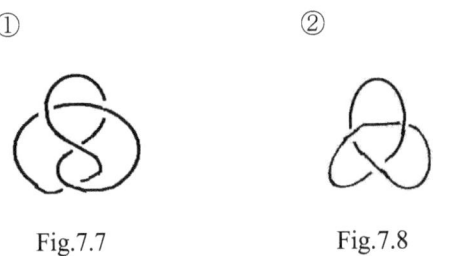

Fig.7.7   Fig.7.8

Problem7.3. Is there a condition that a knot can be untied? Is there a condition that a knot can be deformed into another knot?

Problem7.4. Can the knots ① and ② be untied?   For the untying knot, illustrate the deformation process under consideration of the condition examined in Problem 7.3.

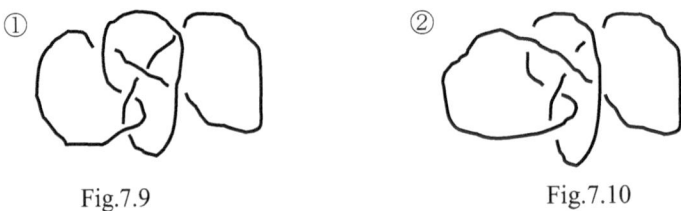

Fig.7.9   Fig.7.10

Problem7.5. Can the knots ① and ② be untied?   For the untying knot, illustrate the deformation process under consideration of the condition examined in Problem 7.3.

① ②

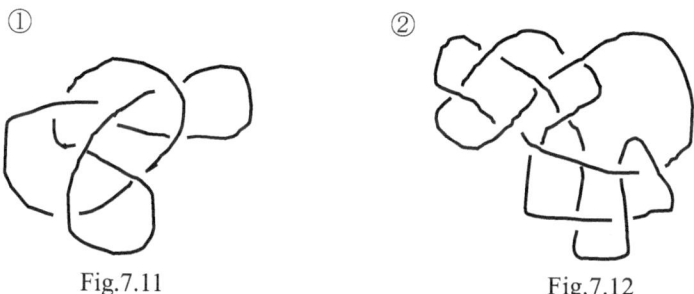

Fig.7.11   Fig.7.12

Two knots are *equivalent or the same knot* if one is deformed into the other by a continuous deformation. In responses of the students through examining Problems above, they showed interests in knots which can be untied, and moved the diagrams little by little. Then they thought of situations of untying knots and noticed the situations in the following notes 1 and 2.

1. A knot diagram in Fig.7.13 can be deformed into a knot diagram in Fig.7.14 where 2 *crossing points* are reduced.

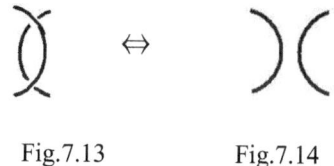

Fig.7.13   Fig.7.14

2. Furthermore, a knot diagram in Fig.7.15 can be deformed into a knot diagram in Fig.7.16 by twisting where one crossing point is reduced.

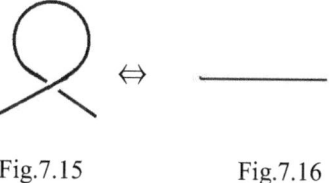

Fig.7.15   Fig.7.16

We did not notice the Reidemeister III although these deformations are called the Reidemeister moves II and I, respectively. The students examined how the knots of Figs.7.13 and 7.17 can be seen from the opposite side. At first, they thought that the opposite of Fig.7.13 is Fig.7.17, and then the opposite of Fig.7.15 is Fig.7.18.

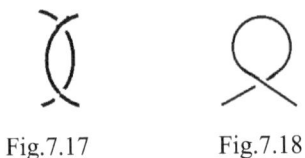

Fig.7.17   Fig.7.18

However, by an actual knot of a string they recognized that they were the same ones.

In responses of the students, the number of crossing points is six in Problem7.4, and seven and twelve in Problem 7.5. The condition on an untying knot in Problem 7.5 was harder than the condition on an untying knot in Problem7.4 by a complexity caused from the number of crossing points.

7.6.2 Reidemeister Moves K. Reidemeister (1893-1971) showed that the equivalence of two knots could be moved into each other by a combination of 3 kinds of moves I, II, III in 1926.

Reidemeister move I :

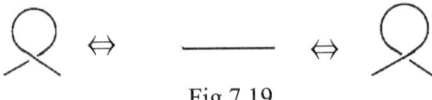

Fig.7.19

Reidemeister move II :

Fig.7.20

Reidemeister move III:

Fig.7.21

When it can be judged that two knots are the same, it is an interesting problem to find where the Reidemeister moves I , II , III are used. However, advanced mathematics on topology are needed in this background. Consider it when you learn an advanced theory from topology to learn knot theory. Knot theory is mathematics to put on the graduate student from the fourth grade in the university. When we consider to untie a knot, we pay attention to a crossing point of a knot diagram to check whether it can be moved.

Problem 7.6. By using the Reidemeister moves, deform ① into ② in Problem 7.2 .

Response of students: Reidemeister move III was hard to be noticed though Reidemeister moves I and II were understood soon in Problem 7.6.

7.6.3 The composition of knot   The knot diagram with 0 crossing point in Fig.7.22 is a trivial knot. The knot diagram with 3 crossing points in Fig.7.23 is called a trefoil knot.

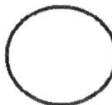   

Fig.7.22:Trivial knot   Fig.7.23: Trefoil knot

The knot diagram with 4 crossing points in Fig7.24 is called the figure-eight knot.

Fig7.24:   Figure-eight knot

The knot diagrams in Figs.7.25 and 7.26 represent the same knot as the figure-eight knot in Fig7.24.

Fig.7.25   Fig.7.26

The composite knot of two knots is a knot obtained from them by removing a small arc from each knot and then connecting the resulting four boundary points with two new arcs. The composite knot up to equivalences is independent of choices of the arcs and determined uniquely from the given two knots. The knot in Fig.7.27 is the composite knot of the figure-eight knot of Fig7.24 and the trefoil knot of Fig.7.23. This operation is similar to the product of positive integers. The composite knot of two non-trivial knots is a non-trivial knot. This is similar to the product of two integers greater than 1. The knot in Fig.7.28 is the composite knot of the trefoil knot of Fig.7.23 and the trivial knot of Fig.7.22, but it equivalent to the trefoil knot of Fig.7.23. This is similar to the product of a positive integer and 1.

Fig.7.27   Fig.7.28

A *prime knot* is a non-trivial knot which is not the composition of two non-trivial knots. It is difficult to judge which knot is a prime knot. For a knot, let n be the minimal number of the crossing points among diagrams of the knot. Then the knot is called a knot of n crossing points. A prime knot of 3 crossing points is only the trefoil knot. A prime knot of 4 crossing points is only the figure-eight knot. Two kinds of prime knots of 5 crossing points are known, and three kinds of prime knots of 6 crossing points are known. Every non-trivial knot is the product of unique prime knots. This is similar to the unique prime decomposition of an integer greater than 1.

7.6.4 Link and linking number A link is a collection of finitely many knots, where the knots are called the components of the link. Thus, a link with one component is a knot. The equivalence of two links is defined in a way similar to the equivalence of two knots. An invariant of a link with 2 components is examined here. The link in Fig.7.29 is called a trivial link with 2 components and the link in Fig.7.30 is called the Hopf link.

Fig.7.29   Fig.7.30

The links in Figs.7.31 and 7.32 are the same link, which is called the Whitehead link. The link in Fig.7.33 is called the Borromean link.

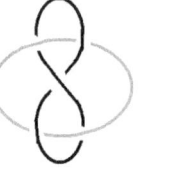

Fig.7.31   Fig.7.32   Fig.7.33

Problem7.7. Can the links in Figs.7.34–7.37 be untied? What condition is needed for untying?

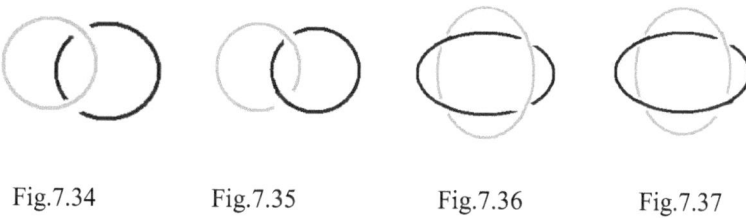

Fig.7.34   Fig.7.35   Fig.7.36   Fig.7.37

Problem7.8. Can the links in Figs.7.38 –7.41 be untied? Is the condition found in Problem7.7 a true condition?

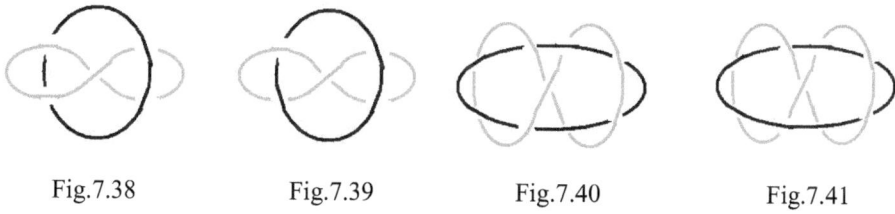

Fig.7.38   Fig.7.39   Fig.7.40   Fig.7.41

Problem7.9.   Can the links in Figs.7.42-7.45 be untied?   Is the condition found in Problem7.7 a true condition?

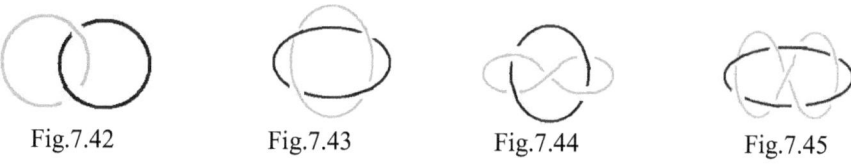

Fig.7.42   Fig.7.43   Fig.7.44   Fig.7.45

In responses of students, these problems were effective for the students to get used to links. When an orientation is given to a link with 2 components, the following sign is attached to a crossing point between the distinct components.

Fig.7.46   Fig.7.47

We do not attach any sign to any crossing point of the same component. The linking number of the oriented link is the half of the sums of all signs.

166

Problem7.10. Compute the linking number of any oriented trivial link with 2 components.

Problem7.11. Compute the linking number of any oriented Hopf link.

Problem7.12. Compute the linking number of any oriented Whitehead link.

As responses of students, in Problem7.10 the linking number is easily seen 0 because there is no crossing point, and in Problems7.11 and 7.12 they examined the following four oriented links because the links are not oriented.

① → →   ② → ←   ③ ← →   ④ ← ←

In Problem7.11:

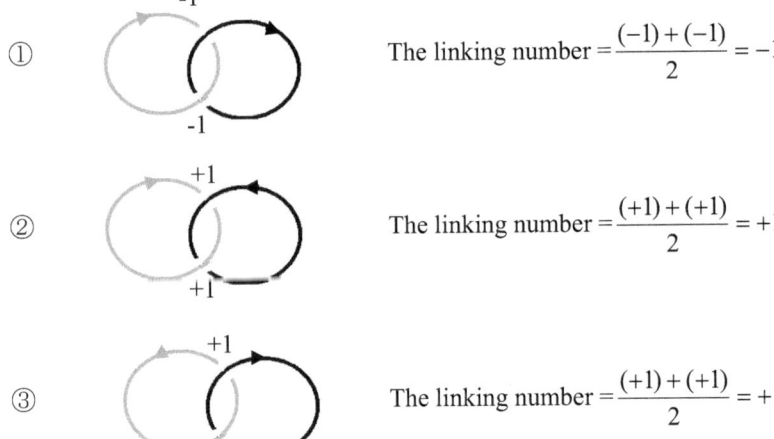

①
The linking number $= \dfrac{(-1)+(-1)}{2} = -1$

②
The linking number $= \dfrac{(+1)+(+1)}{2} = +1$

③
The linking number $= \dfrac{(+1)+(+1)}{2} = +1$

④
The linking number $= \dfrac{(-1)+(-1)}{2} = -1$

In Problem 7.12:

①

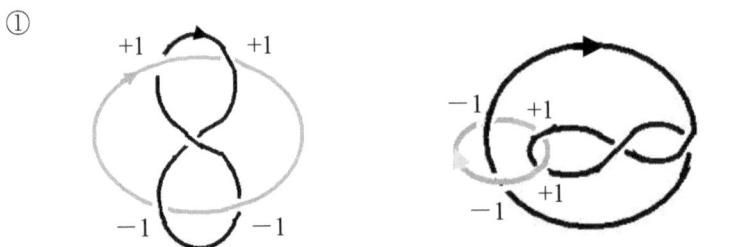

The linking number $= \dfrac{(-1)+(-1)+(+1)+(+1)}{2} = 0$

In the cases of ②③④, the linking number is seen to be 0 in the same way.

In responses of students, they examined a condition that a link can be untied, but did not think of oriented links. The definition of the linking number was given them. It was easy for the students to understand it.

Problem7.13. Find changes of the linking number under the Reidemeister moves Ⅰ, Ⅱ and Ⅲ.

The linking number is unchanged under the Reidemeister moves in the examples described above, but it must be noted that the linking number is an invariant of an oriented link. We see the following: The absolute value of the linking number of a link is independent of a choice of orientations of the components of the link. By the contraposition, we see the following: Two links are not the same if the absolute values of their linking numbers are not equal.

Problem7.14. Compute the linking number of the link (King Solomon knot) in Fig.7.48.

Problem7.15. Compute the linking number of the link in Fig.7.49.

Fig.7.48   Fig.7.49

The linking number of a trivial link is 0. The absolute value of the linking number of Hopf link is 1. These links are different links by not only an appearance but also mathematics. However, because the linking number of the Whitehead link is 0, it is not seen at this stage that the Whitehead link and the trivial link are different links. The trivial link of Fig.7.29 can be deformed into the link in Fig.7.49, so that both the linking numbers are equal to 0. The trivial link of Fig.7.29 cannot be deformed into the Whitehead link though the linking number of Whitehead link of Fig.7.31 is 0. The links of Fig.7.49 and Fig.7.31 are not equivalent because the absolute value of their linking numbers are different. A difference between the trivial link and the Whitehead link cannot be judged by the linking number because their linking numbers are both 0. Thus, we must think of another invariant.

7.6.5 Tri-colorability   A new invariant is examined here.

Problem7.16. Paint the following links. Further, how do you paint them to distinguish between different links?

①   ②   ③

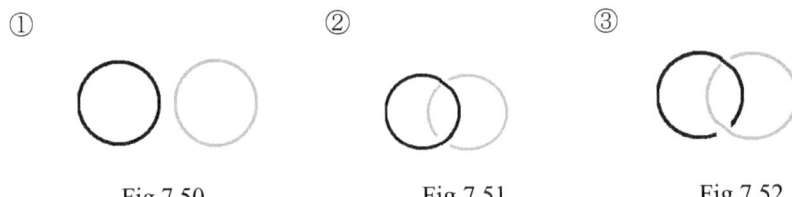

       Fig.7.50   Fig.7.51   Fig.7.52

As responses of students, ① is painted by two colors, and ② is painted by three colors, and ③ is painted by two colors. Since ① and ② are equivalent and not equivalent to ③, if ③ is painted only by one color, then we can see that ③ is different from ① and ② by the coloring condition with three colors or one color at every crossing point.

Problem7.17. Paint the following knots. Further, how do you paint them to distinguish between different knots?

④   ⑤   ⑥

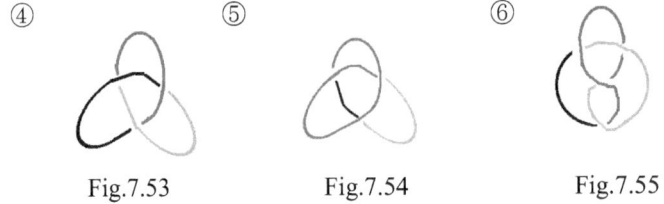

      Fig.7.53   Fig.7.54   Fig.7.55

As responses of students, in Problem 7.17, ④ and ⑤ are painted by three colors and ⑥ is painted by four colors, although we can see from the coloring condition that ⑤ is painted by one color and ④ and ⑥ are painted by three colors, so that ⑤ is different from ④ and ⑥. The following definition comes from considerations of Problems 7.16 and 7.17.

Definition.   A diagram of a knot or a link is tri-colorable when the following conditions (1) - (3) are fulfilled.

(1) Every upper arc of the diagram is painted.

(2) Three colors or one color are used around every crossing point.

(3) Two or three colors are used in the diagram.

The tri-colorability is independent of a choice of diagrams of o a knot or a link. Hence two equivalent knots (or links) have the same tri-colorability. As the contraposition, we see that two equivalent knots (or links) are not the same if one is tri-colorable but the other is not. The trefoil knot is tri-colorable by the argument of Problem7.17. However, the trivial knot is not tri-colorable. It is understood that the trefoil knot is different from the trivial knot.

Problem7.18. Examine the tri-colorability of the knot ① and the link ②.

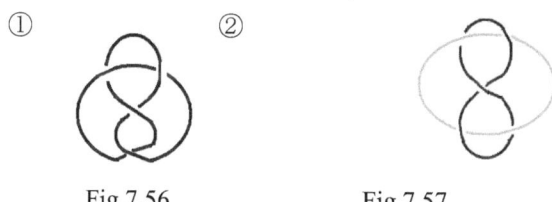

Fig.7.56   Fig.7.57

① and ② are not tri-colorable by the argument of Problem 7.18. The trefoil knot and the figure-eight knot are understood to be different knots as this result. However, we cannot see at this stage that the figure-eight knot is not a trivial knot.

Problem7.19. Specify upper-lower information to the crossing points of the picture of Fig.7.58 to construct a trivial knot and a non-trivial knot. Examine their tri-colorability.

Problem7.20. Examine the tri-colorability of the knot in Fig.7.59.

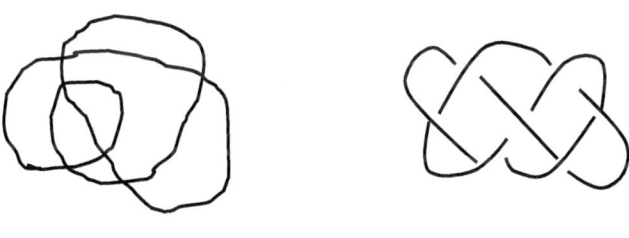

Fig.7.58   Fig.7.59

In a consideration of Problem 7.19, we can obtain examples of a trivial knot and a non-trivial knot which is not tri-colorabile. In a consideration of Problem 7.20. this knot is tri-colorable. Although the linking numbers of the Whitehead link and the trivial link is both 0, it is understood by the tri-colorability that the Whitehead link and the trivial link are different links.

Problem7.21. Examine the tri-colorability under Reidemeister moves Ⅰ, Ⅱ, Ⅲ.

We see that the tri-colorability is unchanged under Reidemeister moves Ⅰ, Ⅱ, Ⅲ, so that the tri-colorability is an invariant of knots and links. It can be judged that the trivial

170

link is different from the Whitehead link, and that the figure-eight knot is different from the trefoil knot. At this stage, we cannot see that the figure-eight knot is different from the trivial knot. We must think another invariant.

7.6.6 Bracket Polynomial A knot invariant is examined further. The *A*-splice is defined by the transformation on diagrams from the crossing arcs in Fig.7.60 into the two arcs in Fig.7.61 fixing the four boundary points ①, ②, ③ and ④. Similarly, the *B*-splice is defined by the transformation on diagrams from the crossing arcs in Fig.7.60 into the two arcs in Fig.7.62 fixing the four boundary points ①, ②, ③ and ④.

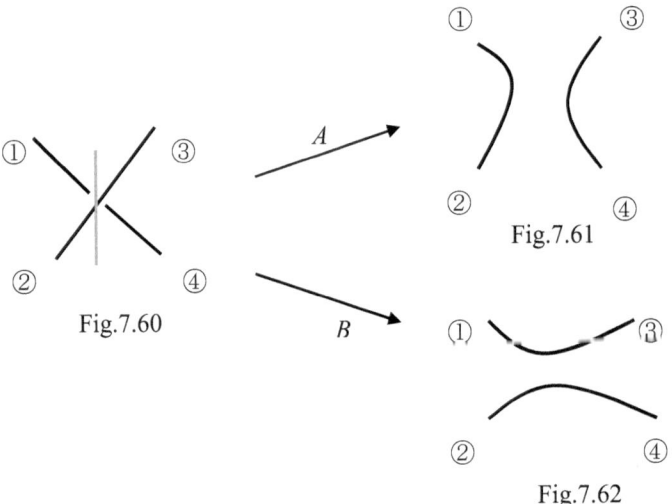

Fig.7.61

Fig.7.60

Fig.7.62

Problem7.22. Illustrate the *A*-splice and *B*-splice of the crossing arcs in Fig.7.63.

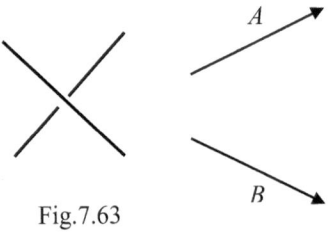

Fig.7.63

We examine the *A*-splice and *B*-splice of the knot diagram *K* with one crossing point in Fig.7.64.

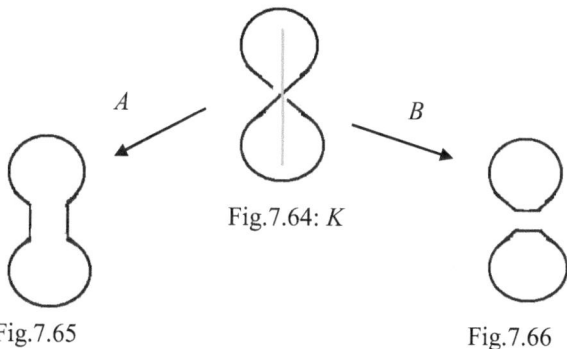

Fig.7.64: K

Fig.7.65   Fig.7.66

The knot diagram K of Fig.7.64 produces the diagram in Fig.7.65 by A-splice and the diagram in Fig.7.66 by B-splice. In this way, we can obtain a diagram without crossing points from any knot or link diagram K by applying A-splice or B-splice to every crossing point. The diagram without crossing point is called a state of the diagram K. This state is shown by the monomial $A^m B^n d^k$ if this state is obtained from K by m times of A-splices and n times of B-splices and this state has k components. We have 2^r states if the diagram K has r crossing points.   The sum of the monomials of the 2^r states of K is defined to be the bracket polynomial of the diagram K and denoted by $B(K)$. The bracket polynomial of knot diagram K in Fig.7.64 is $B(K)=Ad+Bd^2$.

Problem7.23. Compute the bracket polynomials $B(K)$ of the knot diagrams K and K' with one crossing point in Figs.7.67-7.68.

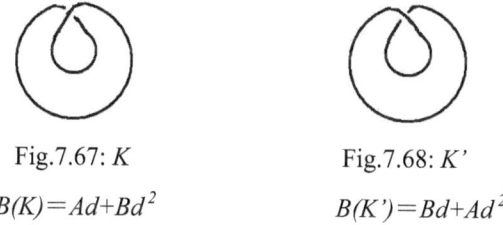

Fig.7.67: K Fig.7.68: K'

$B(K)=Ad+Bd^2$ $B(K')=Bd+Ad^2$

As a response of students, they noticed that the bracket polynomial of the mirror image of a diagram is obtained from the bracket polynomial of the original diagram by exchanging A and B.

Problem7.24. Compute the bracket polynomials of the knot diagrams with 2 crossing points in Figs.7.69-7.70.

172

Fig.7.69   Fig.7.70

In either case, we have $B(K)=A^2d+2ABd^2+B^2d^3$.

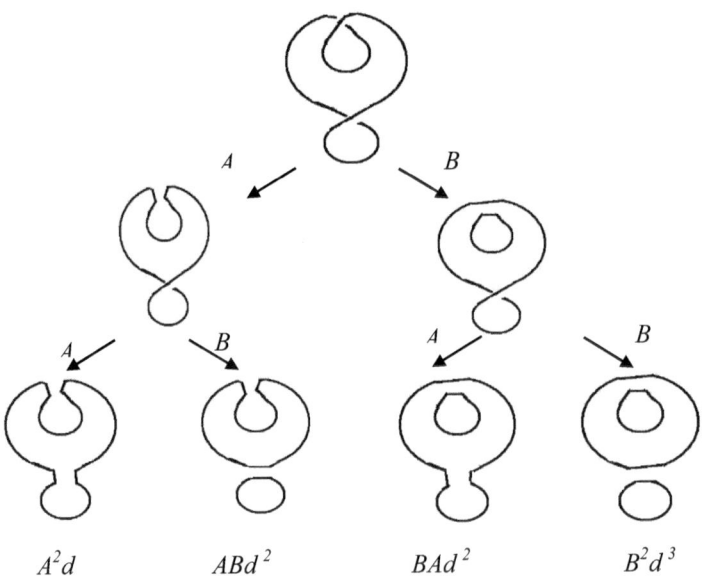

A^2d ABd^2 BAd^2 B^2d^3

Problem7.25. Compute the bracket polynomial $B(K)$ of
the trefoil knot K in Fig7.71.
We have $B(K)=3A^2Bd+(A^3+3AB^2)d^2+B^3d^3$.

Fig7.71

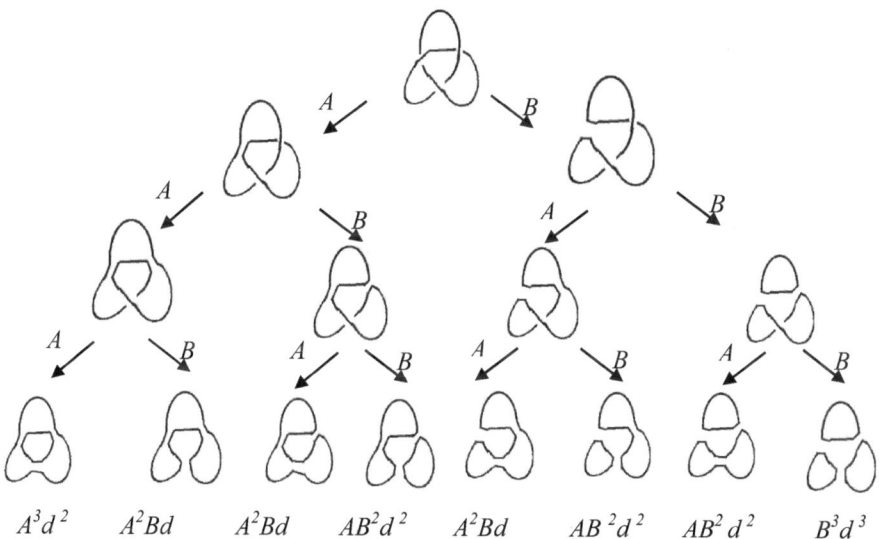

$$A^3d^2 \qquad A^2Bd \qquad A^2Bd \qquad AB^2d^2 \qquad A^2Bd \qquad AB^2d^2 \qquad AB^2d^2 \qquad B^3d^3$$

The following conditions are established from the argument above, where the bracket polynomial of a link diagram K is denoted by $\langle K \rangle$.

Condition (1) For a trivial knot diagram T without crossing point, $\langle T \rangle =1$.

Condition (2)

$$\langle \times \rangle = A \langle)(\rangle + B \langle \asymp \rangle ,$$

$$\langle \times \rangle = A \langle \asymp \rangle + B \langle)(\rangle .$$

For a link diagram $L \cup T$ where L is a link diagram and T is a trivial knot diagram without crossing point splitting from L, we have the following condition.

Condition (3) $\langle L \cup T \rangle = d \langle L \rangle$.

Next, we examine any relations between the unknown variables A , B and d under the Reidemeister moves I , II and III so that the bracket polynomial becomes an invariant of a link K.

Problem 7.26.   Find relations to establish the identity ① by using the conditions (1), (2)and (3).

$$\langle \text{\textcircled{}} \rangle = \langle)(\rangle \quad \cdots ①$$

Problem 7.27. Show that this bracket polynomial is unchanged under the Reidemeister move III by assuming that the identity ① holds .

Problem7.28. Examine whether or not the bracket polynomial is unchanged under Reidemeister move I .

Problem7.29. Compute the bracket polynomials of the following diagrams of knots and links by the relations appearing in Problem 7.26.

① trivial knot   ② trivial link   ③ trivial link of 3 components

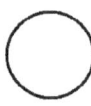  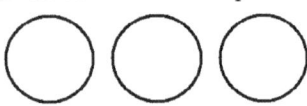

④ knot of Fig.7.67   ⑤ knot of Fig.7.68   ⑥ trefoil knot of Fig.7.71

⑦ knot   ⑧ figure-eight knot

The consideration of Problem 7.26:

$$\langle \widetilde{Q} \rangle = A \langle \widetilde{U} \rangle + B \langle \widetilde{Q} \rangle$$

$$= A \left(A \langle U \rangle + B \langle \}\{ \rangle \right) + B \left(A \langle \widetilde{Q} \rangle + B \langle \widetilde{\Omega} \rangle \right)$$
$$= (A^2 + ABd + B^2) \langle \asymp \rangle + BA \langle)(\rangle .$$

Therefore, $(A^2 + ABd + B^2) \langle \asymp \rangle + BA \langle)(\rangle = \langle)(\rangle .$

From this, we have $A^2 + ABd + B^2 = 0$ and $BA = 1$.

Therefore, $B = \dfrac{1}{A} = A^{-1}$, $d = -A^2 - B^2 = -A^2 - \dfrac{1}{A^2} = -A^2 - A^{-2}$.

The consideration of Problem 7.27:

$$\langle \asymp \rangle = A \langle \asymp \rangle + A^{-1} \langle \asymp \rangle .$$

$$\langle \asymp \rangle = A \langle \asymp \rangle + A^{-1} \langle \asymp \rangle .$$

The bracket polynomials of the left sides are the same because the bracket polynomials of the right sides are the same.

The consideration of Problem 7.28:

$$\langle \, \text{⧜} \, \rangle = A \, \langle \, \text{⧝} \, \rangle + B \, \langle \, \text{⧜} \, \rangle = (A + Bd) \, \langle \, \text{—} \, \rangle = (-A^{-3}) \, \langle \, \text{—} \, \rangle .$$

$$\langle \, \text{⧜} \, \rangle = A \, \langle \, \text{⧜} \, \rangle + B \, \langle \, \text{⧝} \, \rangle = (A\,d + B) \, \langle \, \text{▪} \, \rangle = (-A^{3}) \, \langle \, \text{—} \, \rangle .$$

Thus, the bracket polynomial depends on Reidemeister move I .

The consideration of Problem 7.29:

① The bracket polynomial of T is $\langle T \rangle = 1$ by the condition (1).

② By the condition (3), $\langle T \cup T \rangle = d \, \langle T \rangle = d.$

③ $\langle T \cup T \cup T \rangle = d \, \langle T \cup T \rangle = d^2 \, \langle T \rangle = d^2.$

④ The bracket polynomial of the knot diagram K of Fig.7.67 is $\langle K \rangle = A + Bd = -\dfrac{1}{A^3}$.

⑤ The bracket polynomial of the knot diagram K' of Fig.7.68 is $\langle K' \rangle = B + Ad = -A^3$.

⑥ The bracket polynomial of the trefoil knot diagram of Fig.7.71 is

$$\langle K \rangle = 3A^2 B + (A^3 + 3AB^2)d + B^3 d^2 = \tfrac{1}{A^7} - \tfrac{1}{A^3} - A^5.$$

⑦ The bracket polynomial of knot K is $\langle K \rangle = A^3 + 2B + \left(B^3 + 3A\right)d + Bd^2 = -A^3.$

⑧ The bracket polynomial of figure-eight knot is

$$\langle K \rangle = 5A^2 B^2 + 4\left(A^3 B + AB^3\right)d + \left(A^4 + A^2 B^2 + B^4\right)d^2 = \dfrac{1}{A^8} - \dfrac{1}{A^4} + 1 - A^4 + A^8.$$

Next, we examine when the bracket polynomial becomes an invariant under the Reidemeister move I . The writhe $w\,(L)$ of an oriented knot or link diagram L is the sum of all the crossing sings of L as in Figs.7.46-7.47 where the self-crossings of the components are granted.

L″   L   L′

When we take $w\,(L) = \omega$, we have $w\,(L') = \omega + 1$ and $w\,(L'') = \omega - 1$.

$$\langle L' \rangle = \left(-A^3\right)\langle L \rangle, \quad \langle L'' \rangle = \left(-A^{-3}\right)\langle L \rangle = \left(-A^3\right)^{-1}\langle L \rangle.$$

Taking $-A^3 = \alpha$, we have $\langle L' \rangle = \alpha \langle L \rangle = \alpha^1 \langle L \rangle$ and $\langle L'' \rangle = \alpha^{-1} \langle L \rangle$.

By the way, $\alpha^{-(\omega+1)} \langle L' \rangle = \alpha^{-(\omega+1)} \cdot \alpha^1 \langle L \rangle = \alpha^{-\omega} \langle L \rangle$ and

$$\alpha^{-(\omega-1)} \langle L'' \rangle = \alpha^{-(\omega-1)} \cdot \alpha^{-1} \langle L \rangle = \alpha^{-\omega} \langle L \rangle.$$

The normalized bracket polynomial of an oriented knot or link diagram L is defined as follows:

$$N_L(A) = \left(-A^3 \right)^{-w(L)} \langle L \rangle.$$

Problem7.30. Show that the normalized bracket polynomial is unchanged under the Reidemeister moves I , II and III.

Problem7.31. Compute the normalized bracket polynomials of the diagrams in Problem 30 ④ - ⑧.

The consideration of Problem 7.30:

[Reidemeister move I]

L'' L L'

$\omega(L'') = \omega - 1$ $\omega(L) = \omega$ $\omega(L') = \omega + 1$

$\langle L' \rangle = (-A^3) \langle L \rangle$, $\langle$ $L'' \rangle = (-A^{-3}) \langle L \rangle$.

$$N_{L'}(A) = \left(-A^3 \right)^{-\omega(L')} \langle L' \rangle = \left(-A^3 \right)^{-(\omega+1)} \left(-A^3 \right) \langle L \rangle = \left(-A^3 \right)^{-\omega} \langle L \rangle = N_L(A).$$

In a similar way, we have $N_{L''}(A) = N_L(A)$.

[Reidemeister moves II , III]

The normalized bracket polynomial is unchanged because there is no change in the writhe by these moves.

The consideration of Problem 7.31:

④ $w(K) = -1$ and $N_K(A) = \left(-A^3 \right)^{-(-1)} \left(-\dfrac{1}{A^3} \right) = 1$.

⑤ $w(K')=1$ and $N_{K'}(A)=\left(-A^{3}\right)^{-1}\left(-A^{3}\right)=1$.

⑥ $w(K)=3$ and $N_{K}(A)=\left(-A^{3}\right)^{-3}\left(A^{-7}-A^{-3}-A^{5}\right)=-A^{-16}+A^{-12}+A^{-4}$.

⑦ $w(K)=1$ and $N_{K}(A)=\left(-A^{3}\right)^{-1}\left(-A^{3}\right)=1$.

⑧ $w(K)=0$ and $N_{K}(A)=A^{-8}-A^{-4}+1-A^{4}+A^{8}$.

Thus, it is understood that the figure-eight knot is different from the trivial knot.

As responses of students, they could not prove this fact by the identities, but they were surprised because this fact could be confirmed by another invariant. Moreover, they could construct with much interest a splicing tree of a knot or link diagram to obtain a state, and could understand it well by a result appearing in visual.

7.6.7 Jones polynomial   The Jones polynomial is nothing but the normalized bracket polynomial.

Problem7.32. Show that the following relation can be derived from the condition Ⅱ of the bracket polynomial.

$$A \left\langle \diagdown\!\!\!\!\diagup \right\rangle - A^{-1} \left\langle \diagup\!\!\!\!\diagdown \right\rangle = (A^{2}-A^{-2}) \left\langle \,\right)(\,\right\rangle .$$

The notations L_{+}, L_{-}, L_{0} are used for diagrams which are identical except the parts indicated as follows:

L_{+}	L_{-}	L_{0}
$\omega\,(L_{+})=\omega+1$	$\omega\,(L_{-})=\omega-1$	$\omega\,(L_{0})=\omega$

Problem7.33. Show that the following relation can be derived from the relation of Problem 7.32.

$$-A^{4}N_{L_{+}}\left(A\right)+A^{-4}N_{L_{-}}\left(A\right)=(A^{2}-A^{-2})N_{L_{0}}\left(A\right).$$

The Jones polynomial $V_{L}(x)$ of a knot or link diagram L is defined from the normalized bracket polynomial $N_{L}(A)$ as follows.

$$V_L(x) = N_L\left(x^{\frac{1}{2}}\right) \qquad (A^2 = x).$$

Problem7.34. Show that the following skein relation of Jones polynomial can derived from the relation of Problem 7.33.

$$-x^2 V_{L_+}(x) + \frac{1}{x^2} V_{L_-}(x) = \left(x - \frac{1}{x}\right) V_{L_0}(x).$$

In 1984, V. Jones from New Zealand discovered this polynomial from a research of operator algebras, and was awarded Fields Prize with International Congress of Mathematicians held in Kyoto in 1990. In the calculation of the Jones polynomial, the simplification $V_L(x) = L$ is sometimes used where the skein relation of Jones polynomials is shown as follows:

$$-x^2 L_+ + \frac{1}{x^2} L_- = \left(x - \frac{1}{x}\right) L_0.$$

Problem7.35. Compute the Jones polynomials of the following links.
① Trivial link of 2 components
② Hopf link

The consideration of Problem7. 32:

$$A \langle \,\text{\Large ✕}\, \rangle - A^{-1} \langle \,\text{\Large ✕}\, \rangle$$

$$= A \langle A \langle \,)(\, \rangle + A^{-1} \langle \,\text{\Large ≍}\, \rangle \rangle - A^{-1} \langle A \langle \,\text{\Large ≍}\, \rangle + A^{-1} \langle \,)(\, \rangle \rangle$$

$$= A^2 \langle \,)(\, \rangle + A A^{-1} \langle \,\text{\Large ≍}\, \rangle - A^{-1}A \langle \,\text{\Large ≍}\, \rangle + A^{-2} \langle \,)(\, \rangle$$

$$= (A^2 - A^{-2}) \langle \,)(\, \rangle.$$

The consideration of Problem 7.33:

Let $\alpha = -A^3$. Then $N_L(A) = \left(-A^3\right)^{-\omega(L)} \langle L \rangle = \alpha^{-\omega(L)} \langle L \rangle.$

$$N_{L_+}(A) = \alpha^{-\omega(L_+)} \langle L_+ \rangle = \alpha^{-(\omega+1)} \langle L_+ \rangle = \alpha^{-(\omega+1)} \langle \,\text{\Large ✕}\, \rangle.$$

$$N_{L_-}(A) = \alpha^{-\omega(L_-)} \langle L_- \rangle = \alpha^{-(\omega-1)} \langle L_- \rangle = \alpha^{-(\omega-1)} \langle \,\text{\Large ✕}\, \rangle.$$

$$N_{L_0}(A) = \alpha^{-\omega(L_0)}\langle L_0\rangle = \alpha^{-\omega}\langle L_0\rangle = \alpha^{-\omega}\;\langle\;\supset\subset\;\rangle$$

By Problem7.32, we have

$$A\alpha\;\langle\;\times\;\rangle\;\alpha^{-\omega-1} - A^{-1}\alpha^{-1}\;\langle\;\times\;\rangle\;\alpha^{-\omega+1} = (A^2 - A^{-2})\;\langle\;)(\;\rangle\;\alpha^{-\omega}.$$

From this, we have

$$A\alpha\;\langle\;\times\;\rangle\;\alpha^{-\omega-1} - A^{-1}\alpha^{-1}\;\langle\;\times\;\rangle\;\alpha^{-\omega+1} = (A^2 - A^{-2})\;\langle\;)(\;\rangle\;\alpha^{-\omega}.$$

Thus, $\;A\alpha\,N_{L_+}(A) - A^{-1}\alpha^{-1}\,N_{L_-}(A) = (A^2 - A^{-2})\;N_{L_0}(A).$

Therefore, $-A^4\,N_{L_+}(A) + A^{-4}\,N_{L_-}(A) = (A^2 - A^{-2})\;N_{L_0}(A).$

The consideration of Problem 7.34:

Let $A^2 = x$. By Problem7. 33 and Note stated below, we have

$$-x^2 N_{L_+}(x^{\frac{1}{2}}) + \frac{1}{x^2}N_{L_-}(x^{\frac{1}{2}}) = \left(x - \frac{1}{x}\right)N_{L_0}(x^{\frac{1}{2}}).$$

Since $\;V_L(x) = N_L\left(x^{\frac{1}{2}}\right)\;$, we have $\;-x^2 V_{L_+}(x) + \frac{1}{x^2}V_{L_-}(x) = \left(x - \frac{1}{x}\right)V_{L_0}(x).$

Note. Jones polynomial is usually considered in the variable t by taking $A = t^{-\frac{1}{4}}$, but here it is considered in the variable x by taking $A^2 = x$ by the advice of Kawauchi in this practice because of the simplification of the calculation.

The consideration of Problem 7.35:

① The Jones polynomial of a trivial link with 2 components.

1) The case that the components have different orientations in the plane. From the skein relation of Jones polynomials, we have

$$-x^2 L_+ + \frac{1}{x^2}L_- = \left(x - \frac{1}{x}\right)L_0.$$

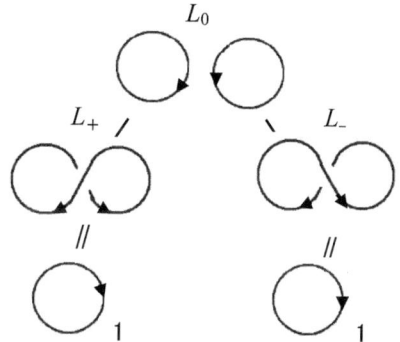

By the resolution trees of the upper right figure and $L_+ = L_- = 1$, we have

$$-x^2 + \frac{1}{x^2} = \left(x - \frac{1}{x}\right)L_0, \qquad \text{Therefore,} \quad L_0 = -\left(x + \frac{1}{x}\right)$$

2) The case that the components has the same orientation in the plane.

We find L_0 below, so that we have $L_0 = -\left(x + \frac{1}{x}\right)$ by the same method as 1).

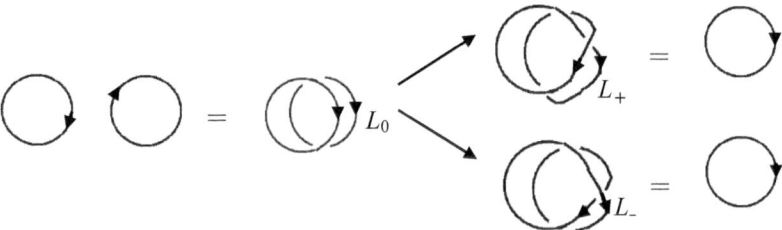

② Jones polynomial of the Hopf link

There are two cases in the Hopf link.

    1)   2)

These links are mutually in the relation of the mirror images and also can be deformed into each other by the Reidemeister moves.

1) The case of oriented Hopf link with a corssing sign +1:
Then all the crossing signs are +1. From the resolution
tree of the right side figure and the skein relation of
Jones polynomials, we have

$$-x^2 L_+ + \frac{1}{x^2}L_- = \left(x - \frac{1}{x}\right)L_0.$$

Since $L_- = -\left(x + \frac{1}{x}\right), L_0 = 1$, we have

$$L_+ = -\frac{1}{x^4}\left(x + \frac{1}{x}\right) - \frac{1}{x^2}\left(x - \frac{1}{x}\right) = -\frac{1}{x^3} - \frac{1}{x^5} - \frac{1}{x} + \frac{1}{x^3}.$$

Therefore, $L_+ = -\left(\frac{1}{x^5} + \frac{1}{x}\right)$

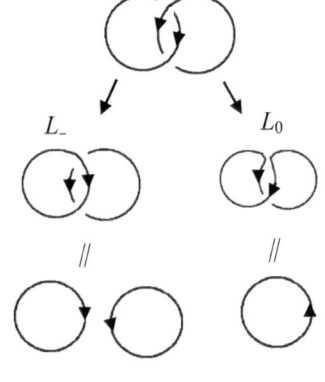

The case of oriented Hopf link with a corssing sign -1:
Then all the crossing signs are -1. From the resolution
tree of the right side figure and the skein relation of
Jones polynomials, we have

$$-x^2 L_+ + \frac{1}{x^2} L_- = \left(x - \frac{1}{x}\right) L_0 .$$

Since $L_+ = -\left(x + \frac{1}{x}\right)$, $L_0 = 1$, we have

$$L_- = -x^4\left(x + \frac{1}{x}\right) + x^2\left(x - \frac{1}{x}\right) = -x^5 - x^3 + x^3 - x .$$

Therefore, $L_- = -\left(x^5 + x\right)$

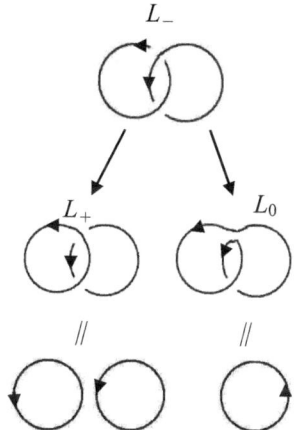

2) The case of oriented Hopf link with a corssing sign -1:
Then all the crossing signs are -1. From the resolution tree of the figure below and the
skein relation of Jones polynomials, we have $L_- = -\left(x^5 + x\right)$.

$$L_+$$

$$L_-$$

$$L_0$$

The case of oriented Hopf link with a corssing sign +1: Then all the crossing signs are
+1. From the resolution tree of the right side figure and the skein relation of Jones

polynomials, we have $L_+ = -\left(\frac{1}{x^5} + \frac{1}{x}\right).$

$$L_+$$

$$L_-$$

$$L_0$$

Problem 7.36 Compute the Jones polynomials of the following knots.

① The trefoil knot.

② The figure-eight knot.

The consideration of Problem 7.36:

① The Jones polynomial of the trefoil knot

 These trefoil knots are mutually in the relation of the mirror images.

1)　　　　　　　2)

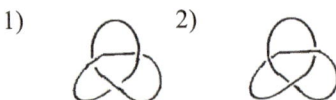

1) Every crossing sign is +1 in any orientation. By the skein relation of Jones polynomials, we have

$$-x^2 L_+ + \frac{1}{x^2} L_- = \left(x - \frac{1}{x}\right) L_0 .$$

Since $L_- = 1$, $L_0 = -\left(\frac{1}{x^5} + \frac{1}{x}\right)$,

we have

$$L_+ = \frac{1}{x^4} + \frac{1}{x^2}\left(x - \frac{1}{x}\right)\left(\frac{1}{x^5} + \frac{1}{x}\right)$$

$$= \frac{1}{x^4} + \frac{1}{x^6} + \frac{1}{x^2} - \frac{1}{x^8} - \frac{1}{x^4} .$$

Therefore, $L_+ = -\frac{1}{x^8} + \frac{1}{x^6} + \frac{1}{x^2} .$

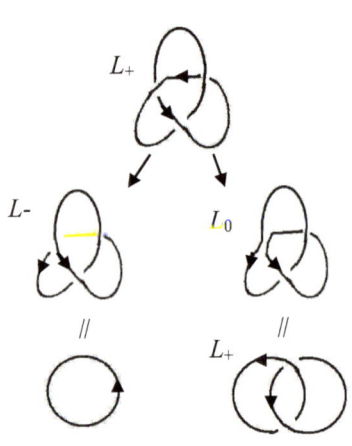

2) Every crossing sign is -1 in any orientation. By the skein relation of Jones polynomials, we have

$$-x^2 L_+ + \frac{1}{x^2} L_- = \left(x - \frac{1}{x}\right) L_0$$

Since $L_+ = 1$, $L_0 = -(x^5 + x)$,

we have

$$L_- = x^4 - x^2\left(x - \frac{1}{x}\right)(x^5 + x)$$

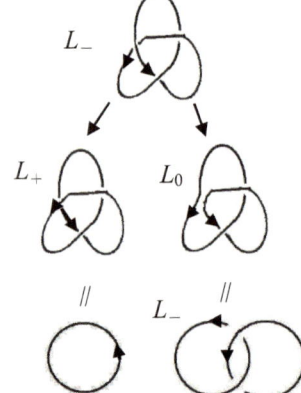

$$= x^4 - x^8 - x^4 + x^6 + x^2.$$

Therefore, $L_- = -x^8 + x^6 + x^2.$

② The Jones polynomial of the figure-eight knot

These two figure-eight knots are mutually in the relation of the mirror images.

1)   2)

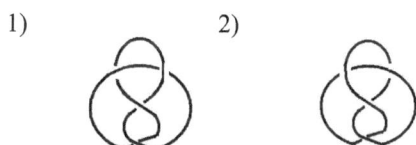

1) From the resolution tree of the right side figure and the skein relation of Jones polynomials, we have

$$-x^2 L_+ + \frac{1}{x^2} L_- = \left(x - \frac{1}{x} \right) L_0.$$

Since $L_+ = 1$, $L_0 = -\left(\frac{1}{x^5} + \frac{1}{x} \right),$

we have

$$L_- = x^4 - x^2 \left(x - \frac{1}{x} \right) \left(\frac{1}{x^5} + \frac{1}{x} \right)$$

$$= x^4 - \frac{1}{x^2} - x^2 + \frac{1}{x^4} + 1.$$

Therefore, $L_- = \frac{1}{x^4} - \frac{1}{x^2} + 1 - x^2 + x^4.$

2) From the resolution tree of the right side figure and the skein relation of Jones polynomials, we have

$$-x^2 L_+ + \frac{1}{x^2} L_- = \left(x - \frac{1}{x} \right) L_0$$

Since $L_- = 1$, $L_0 = -(x^5 + x),$

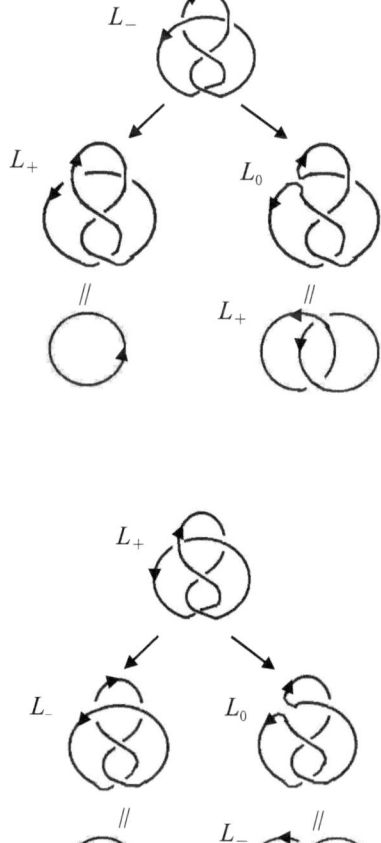

we have

$$L_+ = \frac{1}{x^4} + \frac{1}{x^2}\left(x - \frac{1}{x}\right)\left(x^5 + x\right)$$

$$= \frac{1}{x^4} + x^4 + 1 - x^2 - \frac{1}{x^2}.$$

Therefore, $L_+ = \frac{1}{x^4} - \frac{1}{x^2} + 1 - x^2 + x^4.$

As responses of students, they found that numerator and the denominator of the Jones polynomial are reversed between the Jones polynomials of a link diagram and the mirror image. Though the polynomials could be used, it was difficult to make a resolution tree of knots and links and specially to understand the diagram L_0. Also they could not understand so well how to compute the Jones polynomial, although they felt the solution interesting. Let K^* be the mirror image of a link diagram K.

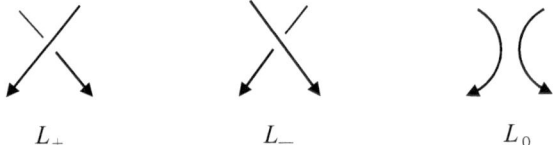

$$L_+ \qquad\qquad L_- \qquad\qquad L_0$$

For the mirror images $(L_+)^*$ of L_+, $(L_-)^*$ of L_- and $(L_0)^*$ of L_0 are respectively regarded as $(L')_-$, $(L')_+$ and $(L')_0$ for a link diagram L'. Hence we see from the skein relation of Jones polynomials the following identity:

$$-x^2 V_{(L_-)^*}(x) + \frac{1}{x^2} V_{(L_+)^*}(x) = \left(x - \frac{1}{x}\right) V_{(L_0)^*}(x).$$

Therefore, $\quad -\frac{1}{x^2} V_{(L_+)^*}(x) + x^2 V_{(L_-)^*}(x) = \left(\frac{1}{x} - x\right) V_{(L_0)^*}(x).$

Therefore, $\quad -\left(\frac{1}{x}\right)^2 V_{(L_+)^*}(x) + \frac{1}{\left(\frac{1}{x}\right)^2} V_{(L_-)^*}(x) = \left(\left(\frac{1}{x}\right) - \frac{1}{\left(\frac{1}{x}\right)}\right) V_{(L_0)^*}(x).$

This shows that Jones polynomial is $V_{L^*}(x) = V_L\left(\frac{1}{x}\right).$

Problem 7.37 Compute the Jones polynomials of the mirror images of a trefoil knot and a figure-eight knot.

The consideration of Problem 7.37:

① The Jones polynomials of the trefoil knot and the mirror image

The mirror image of the trefoil knot R in 1) is the knot R^* in 2).

1)   2)

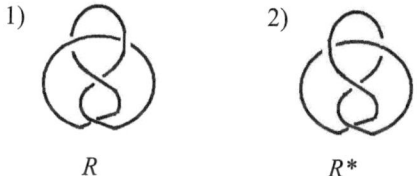

R   R*

The Jones polynomial of R^* is $V_{R^*}(x) = V_R\left(\dfrac{1}{x}\right) = -\dfrac{1}{\left(\dfrac{1}{x}\right)^8} + \dfrac{1}{\left(\dfrac{1}{x}\right)^6} + \dfrac{1}{\left(\dfrac{1}{x}\right)^2}$.

② The Jones polynomials of the figure-eight knot and the mirror image

The mirror image of the figure-eight knot R in 1) is the knot R^* in 2).

1)   2)

R   R*

The Jones polynomial of R^* is $V_{R^*}(x) = V_R\left(\dfrac{1}{x}\right) = \dfrac{1}{\left(\dfrac{1}{x}\right)^4} - \dfrac{1}{\left(\dfrac{1}{x}\right)^2} + 1 - \left(\dfrac{1}{x}\right)^2 + \left(\dfrac{1}{x}\right)^4$.

References

[1] A. Kawauchi et al, Knot Theory (in Japanese), Springer Verlag, Tokyo, 1990. (English expanded version: A Survey of Knot Theory, Birkhäuser Verlag, 1996.)

[2] A. Kawauchi and T. Yanagimoto et al, An Approach to Teaching Knot Theory in School Mathematics for Pupils and Students (in Japanese), Project of Teaching Knot Theory in School Mathematics, Research Report as Educational Action in 21st Century COE Program "Constitution of wide-angle mathematical basis focused on Knots(Osaka City University)" Vol.1 2005, Vol.2 2007, Vol.3 2009.

[3] C. C. Adams, The Knot Book, W.H. Freeman and Company, 1994. (Japanese version

translated by T. Kanenobu, 1998.)

[4] J. Imai, H. Terao and H. Nakamura, What is Invariant? (in Japanese), Spirit of New Mathematics, 2002, BLUE BACKS, Kodannsha Ltd.

[5] L. H. Kauffman, Knots and Physics, World Scientific, 1991. (Japanese version translated by S.Suzuki and A. Kawauchi, 1995.)

[6] S. C. Carlson, Topology of Surfaces, Knots and Manifolds, John Wiley & Sons, Inc., 2001 (Japanese version translated by T. Kanenobu, 2003.)

Index